Nanotechnology in Medicine

Topical Conference at the 2010 AIChE Annual Meeting

Salt Lake City, Utah, USA
7-12 November 2010

ISBN: 978-1-61782-144-8

Printed from e-media with permission by:

Curran Associates, Inc.
57 Morehouse Lane
Red Hook, NY 12571

Some format issues inherent in the e-media version may also appear in this print version.

Printed by Curran Associates, Inc. (2011)

AIChE
3 Park Avenue
New York, NY 10016-5991

Phone: (203) 702-7660
Fax: (203) 775-5177

www.aiche.org

Additional copies of this publication are available from:

Curran Associates, Inc.
57 Morehouse Lane
Red Hook, NY 12571 USA
Phone: 845-758-0400
Fax: 845-758-2634
Email: curran@proceedings.com
Web: www.proceedings.com

TABLE OF CONTENTS

(414a) siRNA Liposome by SCF Technology

Ranjit Thakur, Technical Research & Development, Novartis Pharma AG, CH-4056 Basel, Switzerland

Short interference RNA (now on siRNA) is a new area of biological therapeutics for gene targeting. A major issue associated with gene therapy is delivery of therapeutically active siRNA, which are the mediators of RNAi-induced specific mRNA (messenger RNA) degradation, into target tissue/cell in vivo. One of the most desirable approach is based on delivery systems like liposomes. Lipid based drug delivery system is gaining lot of importance due to self assembling nature of these molecules. Also, lipids can accommodate hydrophilic as well as hydrophobic molecules which give great advantage for sensitive oligonucleotides such as proteins, RNA, siRNA, ds-RNA, antisense etc. There are various methods discussed in literature for liposome formulation such as thin film method using evaporation, sonication method as well as ethanol injection method. However, these techniques suffer from different disadvantages, such as multi-step process, scalability, reproducibility and broad particle size distribution along with others. In this work, Supercritical fluid (SCF) is used for siRNA-lipid nanoparticles formulation. SCF based process is a one step process for siRNA lipoplex formulation with controllable particle size and can be operated at mild conditions. In this process, lipid or mixture of lipids is solubilized in supercritical CO_2 with or without co-solvents and then depressurized jet stream coming out of nozzle mixes with aqueous jet stream of siRNA solution. In current work, nozzle is used to create high velocity stream for better mixing prior to mixing with each other. This work is a first ever attempt to produce nanometer lipid vesicles containing siRNA using SCF method. The major advantage of using this technique is control on the particle size with one step process. This successful attempt is highly encouraging for delivery issues related with siRNA. The size of siRNA-lipid nanoparticles obtained after processing was < 350 nm. Encapsulation of siRNA in lipid bilayers were analyzed using gel electrophoresis. This process can also be used to encapsulate liposome with polymers like PLGA and PEG. These polymers can either be present in aqueous solution or SCF-lipid solution.

See more of this Session: Nanomedicine and Drug Delivery I
See more of this Group/Topical: Topical F: Nanotechnology in Medicine

Encapsulating Emulsions Inside Liposomes for Drug Delivery

Marjan Javadi[1], William Pitt[1], Jonathan Hartley[1] and James R. Lattin[2], (1)Chemical Engineering, Brigham Young University, Provo, UT, (2)Chemical Engineering Dept, Brigham Young University, Provo, UT

Drug delivery of pharmaceutical agents to humans or animals requires understanding of release, absorption, and circulation of the drugs in the body. Some common routes for delivering drugs are through the mouth, skin, nose, vagina, and rectum. Some drugs cannot be delivered through these pathways such as gene- based drugs, vaccines or antibodies. They may not be easily absorbed into the circulation system because of their size or their charge. Injection is usually the preferred way to deliver drugs that are not absorbed well.

One of the main goals of drug delivery is to deliver pharmaceutical agents to a specific location in the body; these drugs must reach specific level of concentration and be maintained in that location to be effective. If it is released in less concentration than is needed the drug will not be effective and if it is released more than that specific level, that might damage the healthy cells. So the concentration of drug in targeted cells is very critical. Drugs can affect both healthy and cancerous cells; sometimes these effects are insignificant and can be avoided, but most of the times for chemotherapeutic drugs the interactions are not negligible and can easily damage healthy cells.

Targeted drug delivery provides a way to avoid the adverse effect of drugs on the healthy cells. This method is used in chemotherapy due to the fact that drug will be delivered to the specific part of the body instead of circulating in the whole body. Additionally in targeted drug delivery less amount of drug is required when it is concentrated at a single site.

Stability and solubility of the pharmaceutical agents are the most important aspects of drugs that should be considered in designing a drug delivery system. The solubility of the drugs depends on temperature, pH, and hydrophobicity.

There are many barriers in the body that protect it from the exposure to any foreign substances, and all drugs will face at least one these barriers which include chemical, physical and physiological barriers. If the drugs cannot pass these barriers, an alternative strategy is to sequester them in vesicles that can penetrate the barriers.

Passive and active targeting are two different methods which are used in targeting drug delivery. In active drug delivery the therapeutic agent is conjugated to some molecule that attaches to the targeted tissue, whereas in passive targeting various carriers without specific targeting ligands are used. Carriers are usually liposomes, emulsions, solid particles and polymeric micelles [1]. Drugs are attached, encapsulated, or entrapped into or on the nanoparticle. In our research group we work with liposomes as a drug carrier. These liposomes are biocompatible and can safely be used as carriers in the human body.

Nanotechnology has solved some of the problems in drug delivery, such as rendering carriers small enough to pass through the physical barriers.

There are 3 general types of nanoparticles: solid, liquid and gas. Nanoemulsions are made of two different phases (hydrophobic and hydrophilic). One is a dispersed phase and the other is the continuous phase (which is usually aqueous), and drug can be carried within the hydrophobic phase or between the phases in drug- soluble oils or lipids. The discrete phase of the emulsion is usually made of hydrocarbon or fluorocarbon. Besides having an important role in targeted drug delivery, perfluorocarbon emulsions have been used as oxygen carriers and also ultrasound contrast agents for imaging [2, 3].

Liposomes are small vesicles whose shells are composed of amphiphilic molecules which can include phospholipids found in the cell membrane. Our liposomes are formed by hydrating phospholipids in an aqueous phase containing the drug. As the spherical liposomes form, drug is encapsulated into the inner aqueous compartment of the liposome. Liposomes have many advantages as drug delivery agents. They are biocompatible and biodegradable and can be functionalized with polyethylene glycol (PEG) to increase the time of circulation.

Ultrasound is used for medical imaging and some therapeutic applications; when applying ultrasound there is no need for surgery. Ultrasound has also been used in targeted drug delivery. The interaction between ultrasound and body may result in thermal or non-thermal effects. When the acoustic energy is absorbed by the tissue, thermal heating occurs. Cavitation, the physical oscillation of bubbles and their collapse, produces some of the non-thermal effects of ultrasound [4].

Applying ultrasound will cause oscillations in the size of the bubbles, and bubbles that have the natural resonant frequency near the ultrasound frequency can reach very high amplitudes of oscillation. Bubble collapse begins to occur when the size of oscillations exceeds twice the resting size. During collapse, extreme temperatures and pressures are generated inside the bubbles, and a shock wave is produced. The bubble usually fragments into smaller bubbles [4].

This collapse which produces a shockwave needs to be controlled because it produces high pressure or temperature which may damage the cells.

Bubbles can increase the effects of ultrasound and act as a drug carrier. There are several commercially available perfluorocarbon gas bubbles, originally developed as ultrasound contrast agents that have been used in ultrasound drug delivery. These include Optison®, Definityis®, and Sonovue®.

A problem with using ultrasound and bubbles for drug delivery is that the bubbles are too large to escape from the circulatory system into the tumor tissue. However, small liposomes, less than 1 micron in diameter, can escape through the leaky capillaries found in many tumors. But liposomes do not contain gas, so they cannot be activated by ultrasound. Our concept is to place nanoemulsions of perfluorocarbon liquids inside liposomes. By using a perfluorocarbon with a boiling temperature near or just above body temperature, we can use ultrasound to cause an expansion of the perfluorocarbon liquid into gas, which will burst, open the liposome and spill the drug. During the low pressure phase of the ultrasound wave, the local pressure drops below the vapor pressure of the perfluorocarbon and causes a phase transition. The challenge is to create liposomes with emulsions and drug inside.

The goal of this particular study was to encapsulate the emulsions of PFC (perfluorocarbon) inside the liposome in one step. Encapsulating the emulsions inside the liposomes can be done either in one step or in 2 steps. In traditional method (2 steps) small liposomes are formed into sheets by adding ethanol. When the ethanol is removed, the sheets will remain intact below the transition temperature of the lipid. Then the emulsions can be added and the mixture heated to reform liposomes that contain emulsions inside. We call liposomes containing encapsulated emulsions “eLiposomes”. We call this a 2-step method because the liposomes are formed twice.

This method takes time, and we wanted to develop a one step method by hydrating original liposomes in presents of emulsion, so the liposomes are formed only once.

Experimental Method

All phospholipids, 1, 2-dipalmitoyl-sn-glycero-3-phosphate (DPPA), 1, 2-dimyristoyl-sn-glycero-3-phosphocholine (DMPC) in chloroform were purchased from Avanti polar lipids, Inc.

(Alabaster, AL). Tetradecafluorohexane and perfluorooctanoic acid (PFOA) was purchased from Sigama- Aldrich (St. Louis, MO). Zonyl® was a gift from DuPont.

Liposomes were made by using DMPC in chloroform (3ml). The chloroform was evaporated on a round bottom flask using the vacuum pump. For making emulsions, DPPA or DPPC (1 ml) was dissolved in 0.5 ml double distilled water (DD, 44°C), Tetradecafluorohexane (0.082 g) was added and the rest of the vial (2ml, Xpertek, Saint Louis, MO) was filled with DD water. The vials were capped in the way that no air bubbles exist in the vial. The vials were shaken in a dental amalgamator (TP-103, GC America, Alsip, IL) for 120 s. Two ml of this emulsion was added to DMPC layers on the flask and the rotovap was used to hydrate the evaporated DMPC with emulsion. This process was done in 30°C. Dynamic light scattering (90 Plus, Particle size analyzer, Brookhaven Instrument Corporation, Holtsville, NY) was used to measure the size of the emulsions, liposomes and eLiposomes.

Results

A

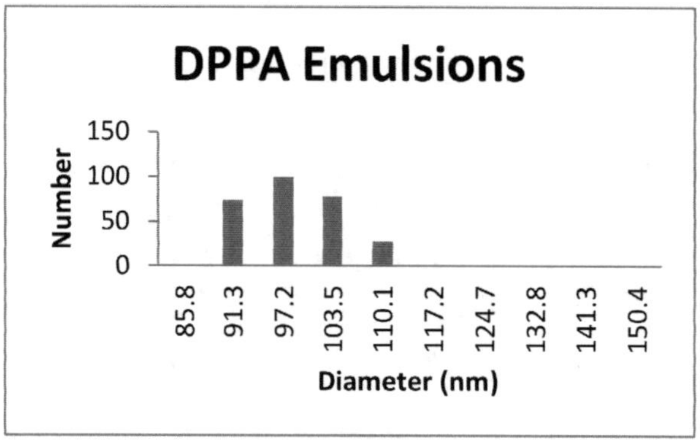

B

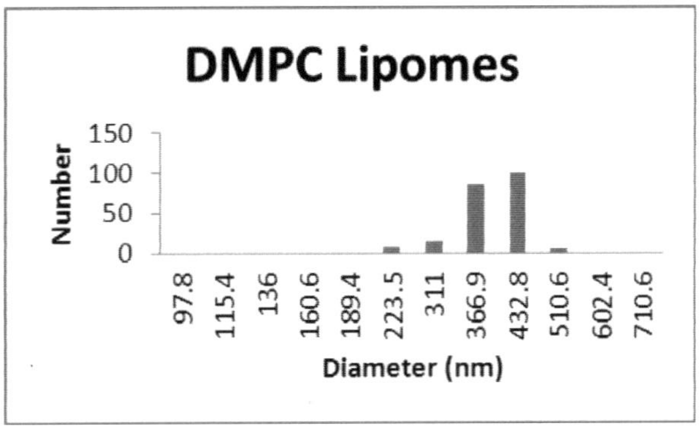

C

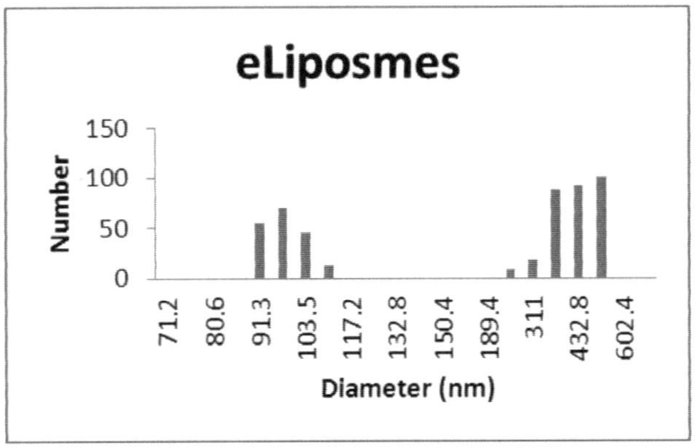

Fig.1. (A) distribution of 1, 2-dipalmitoyl-sn-glycero-3-phosphate (DPPA) emulsions, (B) distribution of 1, 2-dimyristoyl-sn-glycero-3-phosphocholine (DMPC) liposomes, and (C) bimodal distribution of encapsulated emulsions inside the liposomes (eLiposomes).

Figure A shows the distribution of 1, 2-dipalmitoyl-sn-glycero-3-phosphate (DPPA) emulsions, the range size of emulsions is about 90 nm to 120 nm. Figure B is the distribution of 1, 2-dimyristoyl-sn-glycero-3-phosphocholine (DMPC) liposomes in range size of the 200 nm to 500 nm, and in figure C bimodal distribution of the encapsulated emulsions inside the liposomes were displayed. Figure C shows that the number of emulsions in the sample decreased which indicates that some of the emulsions were encapsulated inside the liposomes and made eLiposomes.

We have tried this process with several combinations of various surfactants used for the emulsion and for the liposomes. The table below shows which combination can be used successfully to produce eLiposomes in one-step hydration process. In general, the data indicate that negatively charged emulsion surfactant (phosphoacid) and zwitterionic charged liposomes (phosphocholine) surfactant can be used to produce eLiposomes successfully in one step. Emulsions stabilized with Zonyl or PFOA did not form 1-step eLiposomes.

Emulsion Surfactant	Liposome Surfactant	Successful eLiposome
DMPC	DMPC	No
DMPC	DPPA	No
DMPC	DLPA	No
DSPC	DSPC	No
DSPC	DPPA	No
DSPC	DLPA	No
DPPA	DMPC	Yes
DPPA	DPPA	No
DPPA	DLPA	No
DLPA	DMPC	Yes
DLPA	DPPA	No
DLPA	DLPA	No
Zonyl	DMPC	No
PFOA	DMPC	No

Table 1, different combination of emulsion surfactant with liposomes surfactant. 1, 2-dilauroyl-*sn*-glycero-3-phosphate (sodium salt) (DLPA), 1, 2-distearoyl-*sn*-glycero-3-phosphocholine (DSPC), Perfluorooctanoic acid (PFOA).

This is an interesting result since 2-step eLiposomes can be formed using phosphocholines for the emulsion and the liposome. Although the 1-step formation of eLiposome is much easier than traditional methods, its application is limited to a small subset of emulsion and liposome surfactants.

References:

[1]: Husseini, G. A. and W. G. Pitt (2008). "Micelles and nanoparticles for ultrasonic drug and gene delivery." Advanced Drug Delivery Reviews 60(10): 1137-1152.

[2]: Inayat, M. S., A. C. Bernard, et al. (2006). "Oxygen carriers: A selected review." Transfusion and Apheresis Science 34(1): 25-32.

[3]: Krafft, M. P., A. Chittofrati, et al. (2003). "Emulsions and microemulsions with a fluorocarbon phase." Current Opinion in Colloid & Interface Science 8(3): 251-258.

[4]: Rapoport, N. Y., A. M. Kennedy, et al. (2009). "Controlled and targeted tumor chemotherapy by ultrasound-activated nanoemulsions/microbubbles." Journal of Controlled Release 138(3): 268-276.

[5]: Kisak, E. T., B. Coldren, et al. (2004). "The vesosome - A multicompartment drug delivery vehicle." Current Medicinal Chemistry 11(2): 199-219.

(414c) Multimodal Phase-Shift Nanoemulsions for MRI, Ultrasonography, and Catalysis of Image-Guided Drug Delivery

Natalya Rapoport[1], Kwon-Ho Nam[1], Anne M. Kennedy[2], Allison H. Payne[3], Nicolas Todd[3], Eun-Kee Jeong[3], Dennis L. Parker[3], Jill E. Shea[4] and Courtney Scaife[4], (1)Bioengineering, University of Utah, Salt Lake City, UT, (2)Clinical Radiology, University of Utah, Salt Lake City, UT, (3)Radiology, University of Utah, Salt Lake City, UT, (4)Surgery, University of Utah, Salt Lake City, UT

The paper discusses imaging, acoustic, and therapeutic properties of block copolymer stabilized perfluorocarbon nanoemulsions used as anticancer drug carriers. Three physical factors that trigger droplet-to-bubble transition in liquid emulsions and gels have been identified, namely heat, ultrasound, and injections through fine-gauge needles. Among those listed, ultrasound irradiation was found to be the most efficient factor. Droplet-to-bubble transition and cavitation of formed microbubbles has been identified as a key step in drug release from nanodroplets. In gel matrices, ultrasound-induced droplet-to-bubble transition was catalyzed by pre-existing bubbles. Dependence of droplet vaporization temperatures on droplet size, surface properties, and pressure will be discussed in relation to therapeutic properties. Novel drug-loaded perfluorocarbon nanoemulsions with a unique combination of properties have been developed. Nanodroplets accumulated in tumor volume by passive targeting via extravasation through defective tumor microvasculature. Tumor accumulation of nanodroplets after systemic injections was confirmed by ultrasound and 19F MR imaging. Nanodroplets were susceptible to acoustic droplet vaporization (ADV) and cavitation as manifested by ultrasonography and generation of harmonics in FFT spectra of scattered ultrasound. Introduction of nanodroplets dramatically accelerated and enhanced focused ultrasound (FUS)-induced heating in gel phantoms and in vivo. In a case study, a large breast cancer tumor was completely resolved under the combined action of systemically injected paclitaxel loaded nanodroplets and 1-MHz FUS at 1.74 MPa rarefactional pressure. No tumor recurrence was observed. With the same therapeutic approach, pancreatic tumors were either resolved or substantially regressed. No therapeutic effect from the nanodroplet/ultrasound combination was observed without the drug, indicating that therapeutic effect was caused by the ultrasound-enhanced chemotherapeutic action of the tumor-targeted drug, rather than the mechanical or thermal action of ultrasound itself. The mechanism of drug release in the process of ultrasonically-activated droplet-to-bubble conversion will be discussed. Summarizing, the developed multimodal nanoemulsion drug formulations can be used for the MRI-guided, ultrasound-mediated, targeted tumor chemotherapy.

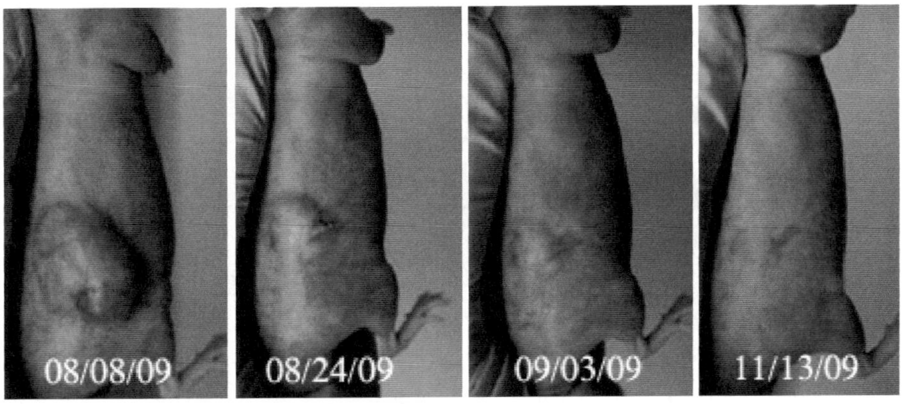

See more of this Session: Nanomedicine and Drug Delivery I
See more of this Group/Topical: Topical F: Nanotechnology in Medicine

(414d) Charge Reversal Liposomes for Cancer Nuclear Drug Delivery

Xinpeng Ma[1], Zhuxian Zhou[1], Bo Zhang[2], Jianbin Tang[3], Maohong Fan[1], Huadong Tang[1], Youqing Shen[3], Maciej Radosz[2], Edward Van Kirk[4] and William Murdoch[4], (1)Department of Chemical and Petroleum Engineering, University of Wyoming, Laramie, WY, (2)University of Wyoming, Laramie, WY, (3)Department of Chemical and Biochemical Engineering, Zhejiang University, Hangzhou, China, (4)Department of Animal Science, University of Wyoming, Laramie, WY

Cationic liposomes have fast cellular uptakes but nonspecific cellular uptakes and are toxic, immediately clear out from the blood, accumulate in lung, Herein, we report liposomes that can undergo negative-to-positive charge-reversal triggered by the solid tumor extracellular- or lysosomal (4-5) acidity for nuclear drug delivery. Doxorubicin loaded in the liposomes is more cytotoxic to tumor cells than the free drug. The liposomes have a long blood circulation time and thus are good for in vivo cancer drug delivery.

See more of this Session: Nanomedicine and Drug Delivery I
See more of this Group/Topical: Topical F: Nanotechnology in Medicine

(414e) Targeted Liposomes with pH-Triggered Leaky Heterogeneities Increase the Therapeutic Potential of Targeted Immunochemotherapy

Amey Bandekar[1], Shrirang Karve[1] and **Stavroula Sofou**[2], (1)Department: Chemical and Biological Engineering, Polytechnic Institute of NYU, Brooklyn, NY, (2)Polytechnic Institute of NYU, Brooklyn, NY

During endocytosis, pH-triggered release of encapsulated therapeutics from delivery carriers may accelerate their intracellular trafficking increasing therapeutic efficacy. To improve the therapeutic potential of targeted immunochemotherapy using anti-HER2/neu liposomal doxorubicin, we exploit the formation of leaky heterogeneities on rigid lipid bilayers to extensively release doxorubicin during endocytosis. We have previously demonstrated that pH-dependent formation of phase-separated lipid heterogeneities on the plane of a bilayer membrane increases the permeability of bilayers when they are composed of lipid pairs with rigid non-matching acyl chain lengths. This was suggested to be due to defective packing among lipids residing at the interfaces of lipid domains. Here we design nanometer-size antiHER2/neu-labeled PEGylated vesicles composed of lipid pairs with longer non-matching acyl chain lengths (n = 18 and 21). In vitro, these vesicles exhibit superior killing efficacy of cancer cells compared to established liposome formulations, and their killing efficacy is similar to the effect of combined free doxorubicin and free antiHER2/neu antibody. Other transport-related properties such as liposome blood circulation times, and specific binding and internalization by cancer cells are unaffected. In vivo, in a murine vascularized breast cancer model, these antiHER2/neu vesicles loaded with doxorubicin exhibit superior tumor control compared to established formulations of liposomal doxorubicin labeled with identical antiHER2/neu antibodies. These results demonstrate the potential of targeted vesicles with pH-triggered leaky heterogeneities to increase the therapeutic potential of targeted immunochemotherapy.

See more of this Session: Nanomedicine and Drug Delivery I
See more of this Group/Topical: Topical F: Nanotechnology in Medicine

(414f) Multicomponent Folate-Targeted Magnetoliposomes: Design, Characterization, and Preliminary in Vitro Hela Cell Studies

Geoffrey D. Bothun, Department of Chemical Engineering, University of Rhode Island, Kingston, RI, Alline Lelis, Chemical Engineering, University of Rhode Island, Kingston, RI and Matthew A. Stoner, Biomedical and Pharmaceutical Sciences, University of Rhode Island, Kingston, RI

Overexpression of folate receptor (FR) is common to many cancer cell lines and provides a means of targeting nanoscale therapeutics. This presentation will describe the design, characterization, and preliminary cytotoxicity results of folate-targeted cationic magnetoliposomes (FTMLs), stabilized by electrostatic and steric repulsion, containing co-encapsulating doxorubicin (DOX, 2 μ M) and anionic superparamagnetic iron oxide (SPIO, 1 μ M) nanoparticles. The FTMLs were designed to achieve combined DOX delivery and hyperthermia under alternating current electromagnetic fields operating at radiofrequencies. By associating the SPIO nanoparticles (5 nm core Fe_2O_3 diamater) with the cationic bilayer through electrostatic attraction, localized nanoparticle heating at the liposome bilayer also provided a mechanism for triggering DOX release. Cryogenic transmission electron microscopy (cryo-TEM) analysis revealed the coexistence of unilamellar and cup-shaped FTML structures, 174 ± 53 nm in diameter, and confirmed that the nanoparticle were encapsulated. This analysis also showed that the FTMLs underwent structural changes after preparation, but retained their cargo. The presence of cup-shaped liposomes with regions of negative bilayer curvature is attributed to bilayer asymmetry and possible domain formation drive by differences in lipid packing parameters. Preliminary in vitro studies conducted on cervical cancer cells (HeLa) confirmed FR targeting and the synergistic effects of DOX delivery and hyperthermia on HeLa cell viability and morphology. While these studies were conducted as a proof-of-concept, to our knowledge this is the first report of FR-targeted magnetoliposome cellular delivery.

See more of this Session: Nanomedicine and Drug Delivery I
See more of this Group/Topical: Topical F: Nanotechnology in Medicine

(414g) Multifunctional PEG-PLL Drug Conjugate Forming Responsive Nanoparticles for Intracellular Drug Delivery

Zhuxian Zhou[1], Jianbin Tang[2], Maohong Fan[1], Huadong Tang[1], Maciej Radosz[3], Edward Van Kirk[4], William J. Murdoch[4] and Youqing Shen Sr.[2], (1)Department of Chemical and Petroleum Engineering, University of Wyoming, Laramie, WY, (2)Department of Chemical and Biochemical Engineering, Zhejiang University, Hangzhou, China, (3)University of Wyoming, Laramie, WY, (4)Department of Animal Science, University of Wyoming, Laramie, WY

poly(ethylene glycol)-b-Poly(L-lysine) was conjugated with different amounts of camptothecin (CPT) via disulfide bonds to form CPT-content controlled glutathione-cleavable micelles in aqueous solution. The micelles were further cross-linked and functionalized with folic acid for enhanced stability in blood circulation and cancer cell-targeting. The multifunctional micelles could enter cancer cells easily through their folate receptor. The cross-links and the disulfide bonds attaching CPT could be cleaved by the high content of glutathione in the cytosol. CPT conjugated nanoparticles had comparable cytotoxicity at low doses to free CPT and higher cytotoxicity at high doses than free CPT, suggesting that the nanoparticles may have potential for improved drug delivery efficacy.

See more of this Session: Nanomedicine and Drug Delivery I
See more of this Group/Topical: Topical F: Nanotechnology in Medicine

(480a) Merging 'Micro' with 'Nano': On-Chip High-Throughput Synthesis of Polymeric Nanoparticles for Cancer Therapy

Pedro M. Valencia, Chemical Engineering, MIT, Cambridge, MA, Minsoung Rhee, Mechanical Engineering, MIT, Cambridge, MA, Robert Langer, Chemical Engineering, MIT, Cammbridge, MA, Omid C. Farokhzad, Department of Anesthesiology, Harvard Medical School, Boston, MA and Rohit Karnik, MIT, Cambridge, MA

Polymeric nanoparticles (NPs) comprised of poly(lactide-co-glycolide)-b-polyethyleneglycol (PLGA-PEG) block copolymers are promising drug delivery vehicles with the advantages of controlled drug release, enhanced stability, and ability to carry thousands of drug molecules per NP [1]. Such NPs can be targeted to differentially deliver drugs to the site of interest in the body, translating to improvement in the therapeutic index of drugs [2]. However, physicochemical properties of NPs such as size, charge, surface chemistry, payload, and targeting ligand density significantly affect the biodistribution of the drug, and the delicate balance between each these properties needs to be experimentally determined and precisely engineered for ultimate in vivo success [3]. Here we report the development of a microfluidic system that enables reproducible, high-throughput preparation of a library of distinct homogeneous, PLGA-PEG NPs. We demonstrate the ability to synthesize and in vitro screen tens of different NP formulations per day by integrating our microfluidic technology with high-throughput flow cytometry (FACS). This makes an optimal system for the translation of nanovehicles to in vivo applications. To the best of our knowledge, this is the first time a microfluidic system is developed for the synthesis and screening of polymeric nanovehicles for drug development applications.

Our microfluidic system, made of PDMS, is composed of a mixing unit and a NP assembly unit. The mixing unit consists on a multi-inlet 2-layer mixer where different precursors such as PLGA-PEG of different MW and charge, PLGA-PEG-folate (targeting agent), drugs, and solvents are mixed at different ratios into a homogenous solution. In the 'assembly' unit, the precursor solution is rapidly mixed with an anti-solvent (i.e. water) using 3D hydrodynamic flow focusing where NPs self-assemble after complete mixing [4]. With this system we can make, in a single step, NPs of different size, charge, surface chemistry, targeting agent density, and drug loading, by simply varying the flow ratios of the streams with different precursors entering the mixing unit.

For the NPs to be effective, it is essential for them to specifically bind to cancerous cells, undergo endocytosis, and release the drug inside the cell [5]. In vitro screening of NPs with varying properties including surface charge, size, targeting ligand density, PEG densities, and chemical composition has the potential to yield valuable insights into the interactions of the NPs at the cellular level. With this aim, we used high-throughput FACS to screen for the uptake of different NP formulations by macrophage cell-lines and cancerous cell-lines as a proof-of-concept. Preliminary data portrays a delicate balance between PEG and folate surface densities for NPs to successfully enter cancer cells while avoiding macrophages. In addition, we demonstrate the capability of our system to combinatorially synthesize and screen NPs against different cell lines.

Our experiments demonstrate a simple yet powerful technology to optimize polymeric nanoparticles in a high-throughput manner by integrating a microfluidic system with high-throughput flow cytometry. This technology forms a robust platform for investigating the effects of different physicochemical properties of NPs on cellular level interactions. This is a clear example where the advantages of microfluidics and nanotechnology are harnessed in a single system.

References 1."Formulation of functionalized PLGA-PEG nanoparticles for in vivo targeted drug delivery," J. Cheng et al, Biomaterials, 28(5), 869 (2007) 2."Targeted nanoparticles for cancer therapy," F. Gu et al, Nano Today, 2(3), 14 (2007) 3."Precise engineering of targeted nanoparticles by using self-assembled biointegrated block copolymers," F. Gu et al, PNAS, 105 (7), 2586 (2008) 4."Microfluidic platform for controlled synthesis of polymeric nanoparticles," R. Karnik et al, Nano Letters, 8 (9), 2906 (2008). 5."Opsonization, biodistribution, and pharmacokinetics of polymeric nanoparticles," D. E. Owens et al, International Journal of Pharmaceutics, 307 (1), 93 (2006)

See more of this Session: Nanomedicine and Drug Delivery II
See more of this Group/Topical: Topical F: Nanotechnology in Medicine

(480b) Fabrication of Highly Uniform Nanoparticles From Recombinant Silk-Elastinlike Protein Polymers for Gene Delivery Applications

Rajasekhar Anumolu, Chemical Engineering, University of Utah, Salt Lake City, UT, Joshua Gustafson, Bioengineering, University of Utah, Salt Lake City, UT, Hamid Ghandehari, Pharmaceutics & Pharmaceutical Chemistry, University of Utah, Salt Lake City, UT and Leonard F. Pease III, Chemical Engineering and Pharmaceutics & Pharmaceutical Chemistry, University of Utah, Salt Lake City, UT

Control of the size and dimensions of silk-elastinlike protein (SELP) polymeric nanoparticles has been demonstrated using an electrospray differential mobility analyzer (ES-DMA) system. The nanoparticles formed as droplets encapsulating the polymer dry. The resulting nanoparticle size distribution depended on the buffer concentration, polymer concentration, and molecular weight of the polymer. The three different polymers used differed in molecular weight based on the number of their silk and elastin units per monomer repeat. The three polymers were SELP-815K, SELP-47K, SELP-415K, with SELP-815K having eight silk motifs, fifteen elastin motifs, and one lysine-substituted elastin motif, etc. The DMA narrowed the size distribution by electrostatically classifying the aerosolized nanoparticles. Transmission electron microscopy images demonstrated that the spherical polymeric nanoparticles are faceted. The faceting appeared to be consistent with a buckling instability, which occurs when the solvent evaporates at the vapor-liquid interface generating a thin hydrogel film that subsequently buckles as additional solvent evaporates. Cylindrical or rod-like particles were also observed presumably caused by formation of polymer stabilized filaments accompanied by a jet instability. The spherical nanoparticles display high uniformity in composition and size, encouraging further development of these nanoparticles as drug and gene delivery carriers.

See more of this Session: Nanomedicine and Drug Delivery II
See more of this Group/Topical: Topical F: Nanotechnology in Medicine

(480c) Formation and Drug Release From Particles Produced Via Flash Nanoprecipitation

Zhengxi Zhu[1], Han Jing[1], Adam Wohl[2], Thomas Hoye[2] and **Christopher W. Macosko**[1], (1)Dept. of Chemical Engineering and Materials Science, University of Minnesota, Minneapolis, MN, (2)Chemistry, University of Minnesota, Minneapolis, MN

Nanoparticles of highly insoluble drugs can be created by flash nanoprecipitation, FNP. Turbulent mixing combined with a polymer stabilizer can produce particles in the 100 nm size range, ideal for passive targeting of tumors and advantageous to increase solubility. We hypothesize that particles form by nucleation of ~10 nm which then aggregate rather than by nucleation and growth. We will report how particle structure controls drug release.

See more of this Session: Nanomedicine and Drug Delivery II
See more of this Group/Topical: Topical F: Nanotechnology in Medicine

(480d) Stabilization of the Nitric Oxide Prodrugs through Incorporation Into PEG-Protected Nanoparticles

Varun Kumar[1], Harinath Chakrapani[2], Sam Y. Hong[3], Anna E. Maciag[4], Joseph E. Saavedra[4], Larry K. Keefer[3] and Robert K. Prud'homme[1], (1)Chemical Engineering, Princeton University, Princeton, NJ, (2)Indian Institute of Science Education and Research, Pune, India, (3)Laboratory of Comparative Carcinogenesis, National Cancer Institute at Frederick, Frederick, MD, (4)Basic Science Program, SAIC - Frederick , Inc., Frederick, MD

We report the stabilization of the nitric oxide (NO) prodrugs and anticancer lead compounds, PABA/NO (O2-{2,4-dinitro-5-[4-(N-methylamino)benzoyloxy]phenyl} 1-(N,N-dimethylamino)diazen-1-ium- 1,2-diolate) and "Double JS-K" 1,5-bis-{1-[(4-ethoxycarbonyl)piperazin-1-yl]diazen-1-ium-1,2-diol-2-ato}-2,4-dinitrobenzene, through their incorporation into polymer-protected nanoparticles. The prodrugs were formulated in block copolymer-stabilized nanoparticles with sizes from 220 to 450 nm by a novel rapid precipitation process. The block copolymers, with polyethylene glycol (PEG) soluble blocks, provide a steric barrier against NO prodrug activation by glutathione. Too rapid activation and NO release has been a major barrier to effective administration of this class of compounds. The nanoparticle stabilized PABA/NO is protected from attack by glutathione as evidenced by a significant increase in time taken for 50% decomposition from 15 min (unformulated) to 5 h (formulated); in the case of Double JS-K, the 50% decomposition time was extended from 4.5 min (unformulated) to 40 min (formulated). The more hydrophobic PABA/NO produced more stable nanoparticles and correspondingly more extended release times in comparison with Double JS-K. The hydrophobic blocks of the polymer were either polystyrene or polylactide. Both blocks produced nanoparticles of approximately the same size and release kinetics. This combination of PEG-protected nanoparticles with sizes appropriate for cancer targeting by enhanced permeation and retention (EPR) and delayed release of NO may afford enhanced therapeutic benefit. Funded by NCI Contract HHSN261200800001E.

See more of this Session: Nanomedicine and Drug Delivery II
See more of this Group/Topical: Topical F: Nanotechnology in Medicine

Anti-Cancer Nanoparticle Synthesis and Characterization

Fan Mei[1], Da-Ren Chen[2] and Yin-Nan Lee[1],
(1)Department of Environmental Science, Brookhaven National Laboratory, Upton, NY,
(2)Department of Energy, Environmental and Chemical Engineering, Washington University, St Louis, MO

For treating the pulmonary disease, the aerosol administration was shown to be more efficient and more controllable in delivering size-selected drug to targeted organ, such as lung, comparing with the traditional drug delivery method. Budesonide, Tarceva, Irresa and resveratrol have been shown to exhibit cancer-preventive activities in preclinical studies and have been proposed to use for the lung-cancer treatments in the animal studies in Washington University in St. Louis. However, the current drug dose estimation is not accurate and the chemical compositions of the produced drug droplets are also unavailable. Thus, a detailed characterization of the drug particles is desirable for improving the drug delivery efficiency through the aerosol administration

In this study, we investigate two drug particles synthesis methods and characterize the size distribution and chemical compositions for the drug particles generated by each method. One of the improved aerosol delivery system is composed with a custom-built atomizer coupled with an efficient solvent remove system. The other one is a compact electrospray coupled with a corona discharger. The generated drug particles size distributions are characterized by a scanning mobility particle sizer (TSI, 3800). The mass ratio between the drugs and the solvent (DMSO, dimethyl sulfoxide) are determined using the mass spectrometer (Aerodyne aerosol mass spectrometer).

For the atomizer method, there is no size effect on the mass ratios of the drugs and the solvent. However, a strong size effect on the drug/solvent mass ratios occurs to the drug particles synthesized by the electrospray, which is believed due to the electrical field induced polarization on the produced drug droplets.

(480f) Doxorubicin- Loaded Albumin Nanoparticles: Formulation and Characterization

Parvin Golbayani, Chemical Engineering, Tennessee Technological University, Cookeville, TN, Soheyla Honary, Pharmaceutics, Mazandaran University of Medical Sciences, Sari, Iran, Mohsen Jahanshahi, Chemical Engineering, Babol Noshirvani University of Technology, Babol, Iran and Pouneh Ebrahimi, High Education Center, Gorgan University of Agricultural Sciences& Natural Resources, Gorgan, Iran

Nanoparticles have the unique property to accumulate at the site of solid tumors and therefore are very suitable for targeted drug delivery1. Different kinds of drugs like pilocarpine, ibuprofen, flubiprafen and ganciclovir have been delivered by entrapping in nanoparticles 2, 3. The objective of the present study was to develop Doxorubicin (Dox) loaded albumin nanoparticles for regional drug release in solid tumors. Bovine serum albumin nanoparticles were prepared by the coacervation method. Drug loading was carried out by two different methods: magnetic stirring the albumin nanoparticles with drug solution as well as shaking the nanoparticles with the drug solution. The entrapment of Dox in albumin nanoparticles was determined to be 39.45% for the magnetic stirring method and 18.58% for the shaking method. In vitro release of the drug was also studied. More than 90% of drug was released in a sustained manner from Dox loaded albumin nanoparticles. To characterize the nanoparticles, particle diameter, zeta potential, and polydispersity were determined by Zetasizer 3600 ZEN Malvern UK. The differential scanning calorimetric method (DSC Perkin Elmer Germany) was used to understand the thermotropic properties and phase transition behavior. Scanning electron microscope methods (2360 Leo Oxford, England) were used to analyze the surface and shape characteristics of the particles after they were coated with gold. The results show that, the size of albumin nanoparticles before drug loading increased from an average of 67 nm to 98 nm for the magnetic stirring method and 106 nm for the shaking method. In addition, the SEM results confirm the uniform shape of particles.

1- A. Arnedo et al., Albumin nanoparticles improved the stability, nuclear accumudation and anticytomegaloviral activity of a phosphodiester oligonucleoide J of Controlled Release 94 pp. 217-227 2004. 2- Das. S. et al., Aspirin loaded albumin nanoparticles by coacervation Trends Biomater. Artif. Organs. 18 pp. 203-211 2005. 3- Merodio M. et al., Ganciclovir – loaded albumin nanoparticles: characterization and in vitro release properties. Europ. J pharm. sci. 12 pp. 251-259 2001.

See more of this Session: Nanomedicine and Drug Delivery II
See more of this Group/Topical: Topical F: Nanotechnology in Medicine

(480g) A New Charge Reversal PCL-Block-Polyhistidine Nanoprticles for Nuclear Targeting Drug Delivery

Erlei Jin[1], Bo Zhang[2], Jianbin Tang[3], Maohong Fan[4], Huadong Tang[4], Maciej Radosz[2], Edward A. Van Kirk[5], William J. Murdoch[5] and Youqing Shen[3], (1)Chemical & Petroleum Engineering, University of Wyoming, Laramie, WY, (2)University of Wyoming, Laramie, WY, (3)Department of Chemical and Biochemical Engineering, Zhejiang University, Hangzhou, China, (4)Department of Chemical and Petroleum Engineering, University of Wyoming, Laramie, WY, (5)Department of Animal Science, University of Wyoming, Laramie, WY

Poly (L-histidine) (polyhis) has an endosomal membrane disruption activity due to its imidazole groups. It would be useful for cytosolic delivery of anticancer drug. However, its low aqueous solubility limits its application. We amidized its imidazole groups and thus substantially enhanced its water solubility. Its block copolymer with PCL could form nanoparticles with a size of around 130 nm. The nanoparticles are negatively charged at pH 7.4 but become positively charged at pH 5.0 due to the hydrolysis of the amides. The nanoparticles can go into nuclei within five hours for effective nuclear drug delivery. CPT and DOX loaded in the nanopartricles had higher cytotoxicity than the free drug.

See more of this Session: Nanomedicine and Drug Delivery II
See more of this Group/Topical: Topical F: Nanotechnology in Medicine

(547a) Multifunctional Rare-Earth Doped Nanoparticles in Encapsulated Albumin Nanocarriers for Tumor Targeting

Dominik J. Naczynski[1], Tamar Andelman[2], Richard E. Riman[2], Charles M. Roth[1] and Prabhas V. Moghe[1], (1)Chemical and Biochemical Engineering, Rutgers University, Piscataway, NJ, (2)Materials Science and Engineering, Rutgers University, Piscataway, NJ

Multifunctional drug probes that can track the delivery of and spatio-temporal responses to localized drugs *in vivo* can serve as powerful diagnostic tools for controlled targeting and therapies. The conventional use of visible-light, excitable fluorophore based probes is limited by poor tissue penetration and autofluorescence, while many inorganic nanoparticles such as quantum dots cause localized toxicity. We have developed a novel family of drug carrier probe nanoparticles that utilize "upconversion fluorescence", which is generated by the excitation of near infrared light and provides highly useful luminescence for biological imaging both in vitro and in vivo. Our approach involves fabrication of water dispersible, biologically targeted, Yb,Er co-doped $NaYF_4$ nanoparticles (REs) by encapsulation in autofluorescent human serum albumin nanocarriers (RE-ANCs). Human serum albumin nanocarriers are biocompatible with many cell types over a large range or concentrations, taken up readily by cells, non-immunogenic, bind reversibly to hydrophobic substances and have numerous functional entities available for conjugating ligands and other moieties. Our nanocomposites can be tuned to have sizes between 75-275 nm, while exhibiting narrow polydispersity and high stability in aqueous solution for over 30 days. The albumin-encapsulation significantly reduced the cytotoxicity of the REs *in vitro* while providing surface amine groups for chemically conjugating targeting ligands. RE-ANCs modified with cyclic RGD rapidly targeted $\alpha V\beta 3$ integrin receptors overexpressed on the U87 glioblastoma cells with minimal targeting of the low integrin expressing A172 cells. Furthermore, we have designed RE-ANCs for the *in vitro* and *in vivo* delivery of a metabotropic glutamate receptor 1 inhibitor, shown to suppress cell proliferation and trigger cell death in melanomas. Outcomes from our research can advance the design and interactions of multifunctional albumin particles as targetable and visualizable drug carriers both *in vitro* and *in vivo* .

See more of this Session: Nanomedicine and Drug Delivery III
See more of this Group/Topical: Topical F: Nanotechnology in Medicine

(547b) Uptake and Clearance of Spherical Gold Nanoparticles in 3D Liver Mimics

Christopher J. Detzel and Padma Rajagopalan, Chemical Engineering, Virginia Polytechnic Institute and State University, Blacksburg, VA

Technological advances in nanotechnology have led the design and synthesis of nanoparticles varying in size and shape that exhibit specific optical, electronic, and chemical properties. Among others, the pharmaceutical, cosmetic, and electronic industries are capitalizing on breakthroughs in nanoparticle technology. Biotechnology is one particular field realizing numerous advances as nanoparticles can be designed with characteristic dimensions resembling those of various biological molecules including bacteria, viruses, or even proteins and DNA. Current biotechnological applications of nanoparticles include imaging and sensing, as well as therapeutic applications in drug delivery and cancer therapy.

With the advent of the nanoparticle industry the prevalence of nanoparticles in everyday life and therefore the risk of human exposure whether it be intentional or inadvertent is increased. With the proven capability of nanoparticles to enter the blood stream after inhalation or ingestion, or directly through injection, evidence exists of nanoparticle accumulation in the liver. However, little is known about the specific uptake and clearance rates of nanoparticles by liver specific cells. In our studies we employ a novel 3D liver mimic culture system to investigate the uptake and clearance rates of 30 nm gold nanoparticles (AuNP).

The 3D liver mimic system is designed to duplicate the in vivo stratified hepatic structure where primary rat hepatocytes are separated from liver sinusoidal endothelial cells (LSECs) by the Space of Disse. In the 3D liver mimic, primary rat LSECs are layered above and separated from primary hepatocytes with the use of a porous self-assembled polyelectrolyte multilayered scaffold. This system enables cellular interactions through autocrine and paracine signaling pathways that are essential for the phenotypic maintenance of each cell type as well as hepatic-specific functions. Our research has shown AuNP uptake by hepatocytes to be directly proportional to both concentration and time during 24 hours of nanoparticle exposure. Results also indicate bare AuNPs are taken up by hepatocytes at a rate 7 times faster than AuNPs coated with a 5K MW PEG polymer. Current studies are underway to quantify the clearance rate of particles from hepatocytes over a 7 day culture period. Studies will also be conducted to understand the transfer and uptake kinetics of AuNPs by each cell type during culture in the 3D liver mimic. Computer tomography and fluorescence measurements are being used to image the sub-cellular localization of the gold nanoparticles. Together, these studies are expected to reveal insights into the interactions between liver tissues and nanoparticles.

See more of this Session: Nanomedicine and Drug Delivery III
See more of this Group/Topical: Topical F: Nanotechnology in Medicine

(547c) Cellular Viability After Electrostatic Deposition of Electrosprayed Nanoparticles

Hedieh Saffari[1], Alexander Malugin[2], Hamid Ghandehari[2] and Leonard F. Pease III[3], (1)Chemical Engineering, University of Utah, Salt Lake City, UT, (2)Pharmaceutics & Pharmaceutical Chemistry, University of Utah, Salt Lake City, UT, (3)Chemical Engineering and Pharmaceutics & Pharmaceutical Chemistry, University of Utah, Salt Lake City, UT

Nanotechnology-based gene and drug delivery strategies suffer from lack of methods to synthesize, characterize, and deliver highly uniform and precisely sized nano-vectors. Here we report the efforts on implementation of a novel method of electrostatic deposition of electrosprayed nanoparticles on cultured cells in vitro. The method employs a nanoparticle aqueous solution, which is first electrosprayed to produce highly charged mists of charged droplets that then can be electrostatically driven towards cellular surface. Multiple factors including intensity of electric field, deposition time, nanoparticles composition,size and concentration could affect cell survival.This study was focused on finding optimal experimental conditions for successful delivery of nanoparticles without compromising viability and morphology of cultured cells.

See more of this Session: Nanomedicine and Drug Delivery III
See more of this Group/Topical: Topical F: Nanotechnology in Medicine

(547d) Responsive and Targeted Nanoparticles for Intracellular Delivery

Weiwei Gao, Brigham and Women's Hospital, Harvard University Medical School, Cambridge, MA, Robert S. Langer, Chemical Enginnering, MIT, Cambridge, MA and Omid C. Farokhzad, Department of Anesthesiology, Harvard Medical School, Boston, MA

We develop a method for prostate-specific small interfering RNA (siRNA) delivery using responsive nanoparticles by functionalization of poly (lactic-co-glycolic acid) (PLGA) and poly (ethylene glycol) (PEG) based copolymer. Here we demonstrate the successful siRNA packaging, targeting and cytoplasm delivery of siRNA into two prostate cancer cell lines: LNCaP and 22RV1. The new delivery system shows the potential to regulate the androgen receptor protein to improve prostate cancer treatment.

Specifically, we chemically modify the PLGA-PEG diblock copolymer backbone to overcome the low siRNA encapsulation ability caused by the hydrophobic nature of PLGA. The PLGA-PEG diblock copolymer is further modified by adding functionalities for preferential escape from the endosome. These chemical modifications eventually lead to the efficient nanoparticle release into the cytoplasm. The nanoparticles are further surface functionalized by conjugation with A10 aptamer, which improves cell uptake via endocytosis by targeting prostate specific membrane antigen on both LNCaP and 22RV1 prostate cancer cells. The siRNA-nanoparticle endosomal escape is visualized by confocal fluorescent microscopy. To assess siRNA delivery efficiency, protein assays were used to demonstrate a decrease in protein expression consistent with activation of the RNAi pathway. In the case of the model GAPDH gene, the KDalert GAPDH assay shows the decrease of GAPDH protein levels. This result is further confirmed by Western blots, which show the decrease of the androgen receptor protein levels relative to loading control. Concurrently, real time PCR using TaqMan Gene Expression Assay probes relevant for the target gene shows relative lower amount of target mRNA relative to beta-actin in the treated cell populations.

Our results demonstrate that the new targeted and responsive PLGA-PEG based nanoparticles are efficient carrier for siRNA delivery and gene silencing in prostate cancer, which may result in favorable development of clinically relevant targeted therapies.

See more of this Session: Nanomedicine and Drug Delivery III
See more of this Group/Topical: Topical F: Nanotechnology in Medicine

(547e) Tuning T Cell Responses by Multi-Functional Nanoparticles

Hong Shen, Department of Chemical Engineering, University of Washington, Seattle, WA, Kenny K. Tran, Chemical Engineering, University of Washington, Seattle, WA, Patrick S. Stayton, Bioengineering Department, University of Washington, Seattle, WA and Anthony J. Convertine, Bioengineering, University of Washington, Seattle, WA

The collaboration of CD4 and CD8 T cells are critical for the establishment of effective immune responses against a variety of diseases. In this study, we fabricate a series of multi-functional nanoparticles in order to shuttle antigen to both class I and class II antigen presentation pathways. We demonstrate the efficiency of class I and class II antigen presentation can be tuned through manipulating the composition of nanoparticles. By engaging both class I and class II antigen presentation pathway, we will be able to effectively coordinate both CD4 and CD8 T cells responses.

See more of this Session: Nanomedicine and Drug Delivery III
See more of this Group/Topical: Topical F: Nanotechnology in Medicine

(547f) Biodegradable Nanoparticles with Sustained Release of Functional siRNA in Skin

Gunilla B. Jacobson[1], Emilio Gonzalez-Gonzalez[2], Ryan Spitler[2], Rajesh Shinde[2], Devin Leake[3], Roger L. Kaspar[4], Christopher H. Contag[2] and Richard N. Zare[1], (1)Department of Chemistry, Stanford University, Stanford, CA, (2)Department of Pediatrics, Stanford University, Stanford, CA, (3)Thermo Fisher Scientific, Dharmacon Products, Lafayette, CO, (4)TransDerm Inc, Santa Cruz, CA

A key challenge in developing RNAi-based therapeutics is efficient delivery of functional short interfering RNA (siRNA) to target cells. To address this need, we have used a supercritical carbon dioxide process to incorporate siRNA in biodegradable polymer nanoparticles (NPs) for in vivo sustained release. By this means we have obtained complete encapsulation of the siRNA with minimal initial burst effect from the surface of the NPs. The slow release of a fluorescently labeled siRNA mimic (siGLO Red) was observed for up to 80 days in vivo after intradermal injection into mouse footpads. In vivo gene silencing experiments were also performed, showing reduction of GFP signal in the epidermis of a reporter transgenic mouse model, which demonstrates that the siRNA retained activity following release from the polymer NPs.

See more of this Session: Nanomedicine and Drug Delivery III
See more of this Group/Topical: Topical F: Nanotechnology in Medicine

COMPARATIVE ANTI-MICROBIAL ACTIVITY OF SUSPENSIONS OF POWDERED COMMERCIAL ZINC OXIDE & WATER-SUSPENDED ZINC OXIDES' NANOPARTICLE ON WIDE-SPECTRUM GRAM POSITIVE & GRAM NEGATIVE BACTERIAL STRAINS

Shilpa J. Newati[1], Sarita Sachdeva[1], Varsha M. Singh[2], Riaz A. Khan*[3]

[1]Department of Biotechnology, [2]Department of Biomedical Engineering, [3]Department of Chemistry, Manav Rachna International University (MRIU), National Capital Region, Aravali Hills, Sector-43, Faridabad, HR 121 001, India

---

ABSTRACT:
The antimicrobial activity of zinc oxide nanoparticles on *Enterobacter aerogenes* was examined. A set of different concentrations of nanoparticles at 50, 75,100 & 125 μg-ml^{-1} were tested on gram negative *Enterobacter* sp. for their concentration dependent antimicrobial effects. A number of other bacteria including *E.coli* & *B.subtilus* were also evaluated against zinc oxide nanoparticles suspension to check their growth. The antimicrobial activity was examined by drawing growth curves using spectrophotometer. The antimicrobial activity was further confirmed by disc diffusion method. The increased concentrations of zinc oxide nanoparticles at μg/ml effectively checked the bacterial growth and increased the inhibition zone diameter. In the presence of zinc oxide nanoparticles, *Enterobacter* growth was significantly reduced while the zinc oxide powder showed lower levels of antimicrobial activity when compared with same dose levels i.e. 50, 75 & 100 μg/ml. Further studies on a mixture of lower levels of antimicrobial drugs and zinc oxide nanoparticles formulation are in progress with a view to effectively curtail the comparatively large dose regimen of the anti-microbial drugs.

ZnO nanoparticles were found to exhibit antibacterial activity against *E.coli, B. subtilis and Enterobact*er aerogenes. The inhibitory effect increases with increase in the concentrations of ZnO nanoparticles. A complete inhibition of microbial growth was achieved at the concentration level of 100μgml^{-1} and higher. A growth study of different microbial cultures was performed in the presence of nanoparticles to observe their effect on the growth profile. This study shows that Zinc nanoparticles have great promise as antimicrobial agent. Concentration and size are two important factors affecting antimicrobial properties of ZnO nanoparticles. The results indicate that ZnO nanoparticles at concentrations higher than 100μgml^{-1} can significantly inhibit the growth of Microbes compared with the control. This antibacterial effect of ZnO nanoparticles observed in this study is consistent with the results recently reported.

KEYWORDS:
Zinc oxide nanoparticles, zinc oxide powder, gram +ve and gram –ve bacteria, anti-microbial activity, ofloxacin.

---

*Corresponding Author: Ph: +91 971 686 4804, kahnriaz@gmail.com, Fax: 0129 4198444 (Prof. Khan)

(607a) Comparative Evaluation of Transitional Cell Carcinoma Treatments

W. T. Godbey and Xiujuan Zhang, Chemical & Biomolecular Engineering, Tulane University, New Orleans, LA

Three drugs were compared for their efficacy in treating murine transitional cell carcinoma of the bladder. Intravesical gene therapy treatments utilizing expression-targeted plasmids, where the murine Cox-2 promoter was used to drive the expression of exogenously inducible forms of caspases 3 and 9, were compared to treatment modalities employing Bacille Calmette-Guérin (BCG) and celecoxib. When administered via lavage, only the gene therapy regimen was found to be effective at restricting tumor progression following a seven-day incubation of tumor tissues. Celecoxib was also administered via the diet to allow for systemic delivery of the drug. The most efficacious celecoxib use tested yielded tumors with masses of (18.3 ± 8.4 mg) versus the gene delivery method, which yielded tumors with masses of (3.6 ± 7.7 mg). The difference was significant (t-test, $n \geq 4$, $P<0.025$). The results showed that the Cox-2 expression-targeted gene therapy system could efficiently bypass the bladder permeability barrier and more effectively inhibit tumor growth and development than either BCG or celecoxib treatments. Long-term data further demonstrated that the gene therapy system could effectively inhibit tumor growth and elongate the life expectancy.

See more of this Session: Nanomedicine and Drug Delivery IV
See more of this Group/Topical: Topical F: Nanotechnology in Medicine

(607b) Formulation, Characterization and Evaluation of Curcumin-Loaded &gamma-Cyclodextrin Liposomal Nanoparticles On Osteosarcoma Cell Lines

Santosh Subhashrao Dhule[1], Patrice Penfornis[2], Trivia P. Frazier[2], Ryan W. Walker[2], Grace Tan[1], Jibao He[3], Radhika Pochampally[4] and Vijay T. John[1], (1)Department of Chemical and Biomolecular Engineering, Tulane University, New Orleans, LA, (2)Center for Gene Therapy, Tulane University Health Sciences Center, New Orleans, LA, (3)Coordinated Instrument Facility, Tulane University, New Orleans, LA, (4)Center for Gene Therapy and Department of Pharmacology, Tulane University Health Sciences Center, New Orleans, LA

Curcumin, the constituent of *Curcuma longa*, has been shown to possess potent anti-neoplastic activity against a number of tumors including prostate, breast and colon cancer. Widespread clinical application of curcumin has been limited due to poor aqueous solubility and low systemic bioavailability. Reports so far indicated extremely low serum levels (22 - 41 ng/ml) of curcumin, even after high amounts (8 g) of oral administration due to its low water solubility (11 ng/ml). Conjugates of curcumin with macromolecules, nano-formulations, cyclodextrins, liposomes and hydrogels are being employed to make the highly hydrophobic drug more water-soluble, thereby increasing its circulation time and bio-availability. In spite of curcumin's effect on broad range of tumors, the anticancer potential of curcumin against osteosarcoma (OS) is not fully explored. Osteosarcoma occurs as a second malignant neoplasm due to a genetic pre-disposition and/or as a consequence of prior cancer therapy. In the current study, we prepared and estimated the potential of liposomal curcumin against osteosarcoma cell lines. We have explored the concept of drug-in-cyclodextrin-in-liposome (DCL) preparations, imaging the complex through cryo transmission electron microscopy. We have also developed a novel tubular liposome delivery technique for curcumin. The presentation will focus mainly on formulation, characterization by cryo-transmission electron microscopy, and the effect of various curcumin formulations on cancer of mesenchymal / epithelial origin and toxicity study on normal human cells that are treated as control. Analysis of cell death pathways used by the OS cell line in response to different curcumin formulations will be described.

See more of this Session: Nanomedicine and Drug Delivery IV
See more of this Group/Topical: Topical F: Nanotechnology in Medicine

(607c) Sustained, Targeted Intraocular Delivery of Therapeutics for the Treatment of Age-Related Macular Degeneration

Rangaramanujam M. Kannan, Chemical Engineering and Materials Science/NICHD Perinatology Research Branch, Wayne State University, Detroit, MI, Bharath Raja Guru, Chemical Engineering, Wayne State University, Detroit, MI, Raymond Iezzi, Ophthalmology and Kresge Eye Institute, Vitreoretinal Service, Department of Ophthalmology, Rochester, MN and Manoj K. Mishra, Department of Chemical Engineering and Material Sciences, Wayne State University, Detroit, MI

Neurodegenerative diseases such as retinitis pigmentosa and age related macular degeneration are due to degeneration of photoreceptor cells of retina. Though some of the neuroprotective drugs like fluocinolone acetonide (FA) are available for treatment, it is difficult to deliver therapeutic amount of drug to the outer retina for a sustained period. We have discovered that PAMAM dendrimers (generation 4) with neutral surface groups (-OH), are able to selectively localize in the cells associated with the neuroinflammation process, upon intravitreal delivery. Building on this, we have developed polyamidoamine (PAMAM) dendrimer based nanodevices for sustained delivery of neuroprotectants, such as fluocinolone acetonide. The drug release from the ester-bonded conjugates reveal that the release shows a nearly zero-order profile over a two month period. The in vivo evaluation in a RCS rat model of retinal degeneration was performed using electroretinography (ERG) measurements of the overall health of the retina, and at the tissue level using microglial and activated microglial, and outer nuclear cell counts. The dendrimer-based formulation was significantly more effective than the free drug, and a drug release implant. One injection of 0.1 microgram of the drug containing 0.7 microgram of the dendrimer was able to arrest retinal degeneration, and preserve cell counts for an entire month. The mechanism of dendrimer-based sustained intracellular delivery, results of in vivo efficacy of drug-dendrimer nanodevices, biodistribution of fluorescently-labeled dendrimers in RCS rats with the disease, and healthy rats will be presented.

Reference:

1. Dendrimer-containing particles for sustained release of compounds, R. Kannan, R. Iezzi, S. Kannan, US patent filed 10/5/07 (Application #60/997987)/International patent filed Oct 2008 (application #, PCT/US2008/078988). Regular patent filed in US, Canada, European Union, UK and India (Apr. 2010)

See more of this Session: Nanomedicine and Drug Delivery IV
See more of this Group/Topical: Topical F: Nanotechnology in Medicine

(607d) Synthetic Zwitterlation of Enzymatic Proteins for Stability and Retained Enzymatic Kinetics

Andrew J. Keefe and Shoayi Jiang, Chemical Engineering, University of Washington, Seattle, WA

A poly(zwitterion)-protein conjugate has been made to provide protein stability in human serum, high concentration urea, and at high temperatures. A new synthetic method has been developed to produce a N-hydroxysuccinimide terminated zwitterionic polymer for use in protein conjugation. Good control and PDI's are shown due to its synthesis in organic solvents followed by post modification into its zwitterionic structure. Alpha-chymotrypsin was chosen as a model enzyme for conjugation to the zwitterionic polymer, poly(carboxybetaine) (pCB). Poly(ethylene glycol) (PEG), a well known and widely used protein stabilizing polymer, was used as a comparison. We have shown that not only do pCB enzyme conjugates posses show stability under the tested conditions, but do so without adjusting their kinetics. Polymer density was also adjusted (polymers/protein) and shown how higher density polymers provide increased stability with both polymers. Commonly, as seen with PEG, polymer inhibition is popularly known for increasing the dissociation constant of the enzyme substrate complex. We hypothesize that this is due to the presence of the highly solvated polymer preventing diffusion of substrate into the active site. pCB, as a poly(osmolite) of betaine, interacts more with the protein surface and less with the surrounding water, therefore stabilizing the structure of the protein, without inhibiting substrate diffusion. This provides evidence for use of pCB polymers to provide stability to protein conjugates for use in the biomedical industry. This provides evidence for using pCB as a more effective polymer conjugate for enzyme and receptor binding proteins.

See more of this Session: Nanomedicine and Drug Delivery IV
See more of this Group/Topical: Topical F: Nanotechnology in Medicine

(607e) Functionalized Alumina Particles for pH-Responsive Drug Delivery

Brad Gordon[1], Daniel Lim[2], Ezinne Achinivu[3], Charles E. Luckett[2], Sheryl H. Ehrman[4] and Douglas S. English[5], (1)Department of Chemical and Biomolecular Engineering, University of Maryland, College Park, MD, (2)Department of Chemistry and Biochemistry, University of Maryland, College Park, MD, (3)Chemical and Biomolecular Engineering, University of Maryland, College Park, MD, (4) Department of Chemical and Biomolecular Engineering, Univeristy of Maryland, College Park, MD 20742, USA, University of Maryland, College Park, MD, (5)Department of Chemistry and Biochemistry, University of Maryland, College Park, MD 20742, USA, University of Maryland, College Park, MD

pH-responsive particles for anti-cancer drug delivery were created using alumina particles functionalized with silane coupling agents. Three types of silane coupling agents with varying chain lengths and terminal groups were used. XPS analysis was done to confirm surface coupling of the silanes onto the particles. In order to observe the effects of the presence of various terminal groups the uptake and release of the positively charged anti-cancer drug Doxorubicin (Dox) and the negatively charged dye sulforhodamine 101 (SR101) were studied. Uptake experiments were performed while incubating the particles in a dilute solution of each compound. Afterwards, release experiments were carried out at 37 ° C in phosphate buffered saline solutions (0.16 M) at pH values of 5.0 and 7.4. The loading ability observed during the uptake experiments was found to be directly proportional to the silane concentration used during functionalization. Zeta potential measurements revealed that the charge on the particles was directly proportional to both the silane type as well as the silane concentration used during functionalization. Long chain silanes were identified to have the best response to pH and showed a two-fold difference in the relative amounts released when comparing trials run at pH 5.0 and pH 7.4. These results suggest that charge interactions between the loaded compound and the terminal silane groups influence not only the rates of uptake and release, but also the pH-responsive character of these particles.

See more of this Session: Nanomedicine and Drug Delivery IV
See more of this Group/Topical: Topical F: Nanotechnology in Medicine

(607f) Cell Type-Dependent Uptake of PEGylated Nanoparticles

Kenny K. Tran, Chemical Engineering, University of Washington, Seattle, WA, Alyssa Sheih, Bioengineering, University of Washington, Seattle, WA and Hong Shen, Department of Chemical Engineering, University of Washington, Seattle, WA

Poly(ethylene glycol) (PEG) has widely been used to endow "stealth" abilities for cell-targeted delivery of nanoparticles. However, few studies have examined the effect of PEG density on particle surfaces on the uptake through various endocytic pathways. Furthermore, different cell types may use different mechanisms to internalize nanoparticles, thus potentially reducing the "stealthy ability" of PEGylation. In this study, we systematically varied PEG density on polystyrene nanoparticles. We examined the uptake of PEGylated nanoparticles in a variety of cell lines derived from various tissues. Through the use of inhibitors, we demonstrated that PEGylation of nanoparticles was only able to reduce the uptake through receptor-mediated endocytosis. Our study provides insights into alternative strategies besides the prevention of protein adsorption on nanoparticles are required to increase the stealthy ability of nanoparticles.

See more of this Session: Nanomedicine and Drug Delivery IV
See more of this Group/Topical: Topical F: Nanotechnology in Medicine

(607g) The Sound of Silence. Multiscale Molecular Simulations and Experiments in Developing Nanocarrier/Nucleic Acid Systems

Sabrina Pricl, Mose-Dicamp, University of Trieste, Trieste, Italy, Paola Posocco, University of Trieste, Trieste, Italy, Maurizio Fermeglia, Molecular Simulation Engineering Laboratory, Department of Chemical Engineering, University of Trieste, Trieste, Italy, Kostantinos Karatasos, Departiment of Chemical Engineering, University of Thessaloniki, Greece, Ling Peng, Department of Chemistry, CNRS, Marseille, France and Dave K. Smith, Department of Chemistry, University of York, United Kingdom

Although the use of DNA and/or potent, sequence-specific small interfering RNAs (siRNAs) to suppress expression of specific gene transcripts was originally a useful technique for probing gene functions in vitro, their successful application in vivo in animal models against a spectrum of diseases, including cancer, has spurred interest in developing this approach for nucleic acid-based therapeutics. However, there are still significant obstacles to be overcome before these models can be used in the clinic as, for instance, anticancer agents. Perhaps, foremost among these is the issue of delivery. The in vivo use of DNA/siRNA effective against cancer hinges on the availability of a delivery vehicle that can be systematically administered to reach both primary and metastatic tumor cells. Moreover, because sufficient intact, functional genetic material must be delivered into cells to reach an effective intracellular concentration, and to limit potential side effects due to a randomized, general transfection of normal, non-target tissues, it is also crucial to develop means of directing such a delivery vehicle specifically to the target cells. Injected nano-scale drug delivery systems, or nanovectors, are ideal candidates to provide breakthrough solutions to the time-honored problem of optimizing therapeutic index for a treatment. Even modest amounts of progress towards this goal have historically engendered substantial benefits across multiple fields of medicine, with the translability from, for example, a subfield of oncology to a field as distant as the treatment of infectious diseases being granted by the fact that the progresses had a single common denominator in the underlying technological platform. In this study, we combined multiscale molecular simulations and experimental approaches to define mode and molecular requirements of the interaction of nucleic acid-based therapeutics and dendrimer/dendron-based delivery agents. In details, the interaction of DNA and different siRNA sequences were complexed with different type of dendrimeric nanovectors. Their binding processes were characterized by a combination of experimental and molecular simulations techniques, and the effects of siRNA sequences, dendrimer generations, and other molecular parameters such as nanocarrier flexibility upon binding, multivalency effects, microstructure formation and phase diagram, complex stability, transfection ability, and silencing efficiency of these systems were rationalized on the basis of the resulting evidences. This type of investigation can provide valuable information to devise optimal delivery systems that would increase the efficacy of DNA/siRNA therapeutics in cells and laboratory animals and move them toward clinical applications.

Functional Magnetic Nanoparticles for Efficient Malaria DNA Vaccine Delivery

Fatin M. Nawwab Al-Deen[1], Jenny Ho[1], Cordelia Selomulya[1*], Charles Ma[2] and Ross Coppel[2]

1 Department of Chemical Engineering, Clayton Campus, Monash University, Australia

2 Department of Microbiology, Clayton Campus, Monash University, Australia

Abstract

Efficient DNA transfection is crucial for DNA vaccine delivery, as low efficacy is often observed in many *in vitro* and *in vivo* studies. To overcome the problem due to cell barriers, the gene vector can be associated with narrow size superparamagnetic nanoparticles, followed by application of an external magnetic field to target the vector to its desired direction. This technique has been proven to result in excellent cell transfection level and to speed up the duration of the process from hours to a few minutes and minimize the vector dose. Thus, the undesired toxicity of magnetic nanoparticles is also reduced. The objective of this study is to improve malaria DNA vaccine delivery with magnetofection by using the membrane associated 19-kDa COOH terminal fragment of merozoite surface protein ($MSP1_{19}$) that plays a critical role in *plasmodium* immunity and is now a leading malaria vaccine candidate.

Introduction

Malaria is one of the most prevalent and devastating of all human parasitic diseases, and it exacts a heavy toll of death and illness in the world. The greatest burden of malaria is borne by children and pregnant women in developing countries. According to the World Health Organization, between 300 and 500 million people contract malaria with1.2 to 2.7 million deaths every year [1-2].Treatment and control have become more difficult with the spread of drug-resistant strains of parasites and insecticide-resistant strains of mosquitoes that carry the parasites. Thus, there is an urgent need for an affordable and effective vaccine for malaria that can promote the fight against this deadly disease. One target growth-inhibitory antibody is the membrane associated 19-kDa COOH terminal fragment of merozoite surface protein ($MSP1_{19}$) which plays a crucial role in *plasmodium* immunity and is now a leading malaria vaccine (Diggs, Ballou et al. 1993; Good, Kaslow et al. 1998). A primary problem limiting the use of pDNA vaccines is the efficiency of pDNA uptake and expression by the antigen presenting cells. Along with the development of nanotechnology, scientists have used materials for gene delivery. For ideal gene delivery, the gene vector should satisfy the requirements of successful gene delivery, safety to the host, resistance to the enzymatic degradation and specific targeting ability into vivo. Recently, Superparamagnetic Iron oxide Nanoparticles (SPIONs) have attracted significant interest for use in gene delivery applications because of low toxicity and low cost production, while biological materials can be easily immobilized on the magnetic particles and directed to the target place by using an external magnet. PEI polymer is well known to form cationic complexes to interact with non-specific, negatively charged DNA and enter the cell via endocytosis. PEI exhibits a high buffering capacity and offers an extensive positive charge density [3-4]. Thus, using PEI-coated SPIONs for gene delivery can further increase the efficiency of gene delivery in *vitro* and also can be physically directed to the specific target site *in vivo*. Malaria vaccines are under developed and as yet, the best technique for enhancing the efficiency of nonviral vector-mediated gene expression of malaria vaccine *in vitro* and *in vivo* still needs further investigation. This is the first time that magnetofection technique is used to deliver malaria DNA vaccine *in vitro*. Successful outcomes would mean a beneficial technique for *in vivo* malaria gene delivery and a new tool for research to possibly overcome many of the problems of *in vivo* malaria gene transfer.

Materials and Methods

Aqueous dispersion of superparamagnetic Fe_3O_4 nanoparticles (SPIONs) with narrow size distribution around 10 nm ±3nm were synthesized via co-precipitation method using ferric and ferrous salts with a Fe^{3+}/Fe^{2+} mole ratio equal to 2in the presence of trisodium citrate ($C_6H_5Na_3O_7.2H_2O$) as an electrostatic stabilizer. The synthesized nanoparticles were characterized by various analytical techniques. Zetasizer Nano ZS (Malvern Instruments Ltd., UK) was used to determine the hydrodynamic diameter in suspension and the zeta potential; TEM and SEM were used for size determination and to observe the surface morphology; X-ray powder diffraction (XRD) (Philips 1140 PW diffractometer with nickel-filtered Cu Ka radiation (λ = 1.5405 A°)) was used to determine crystallinity and phase of the iron oxide particles. The response to an external magnetic field was measured using a vibrating sample magnetometer (VSM, RIKEN DENSHI) in a field of up to 15 kOe at room temperature.

Among the polycations presently used for gene delivery, PEI (polyethylenimine) has emerged as a very interesting candidate due to its potential for endosomal escape. The magnetic nanoparticles were coated with PEI at different PEI: SPIONs mass ratio. The plasmid DNA vector encoding malaria geneMSP1-$_{19}$ was condensed on the nanoparticles with different N/P (molar ratio of PEI nitrogen to DNA phosphate) under two pH conditions 4.0 and 7.0.Agarose electrophoresis (0.8% agarose) in 1× Tris acetate –EDTA buffer for 90 min at 60 v was used to determine DNA binding to the PEI polymer. Complexes of SPIONs/PEI with MSP1-$_{19}$ plasmid were formed at N/P ratios of 0.5, 1, 2, 5, 7, 10, 15, 20, 25, and 30 with SPIONs: PEI ratio of 10 at pH 4.0 and 7.0.

Evaluation of the cell toxicity was performed by MTT assay in African Green Monkey Kidney Fibroblast Cells (COS-7 cell line).The COS-7 cells were seeded in a 96 well plate at a density of 2×10^4cells /well at 37 °C in 5% CO_2 atmosphere. After 24 h of culturing, different complexes were prepared to test their toxicity in the COS-7cell line. COS-7 cell lines transfected with polyplexes prepared by using SPIONs-PEI-DNA polyplexes which were prepared previously under two pH conditions 4.0 and 7.0 with the different N/P ratios (1-30), DNA/Lipofectamine 2000 complex, and three different concentration of naked SPIONs (0.1,0.5 and 1) mg/ml. The samples were all seeded for further 24 h incubation. The control well was a culture medium with no particles.

For gene transfection, COS-7cell lines were cultivated in the complete RPMI 1640 medium containing 10% fetal calf serum, 2 mM of L-glutamine, 100 μml^{-1} streptomycin and 100 μml^{-1} of penicillin. COS-7cell line were seeded in 6-well at a density of 2×10^5 cell/well and cultured for 12h at 37°C and 5% CO_2 before transfection. SPIONs/PEI/DNA polyplexes were prepared using different N/P ratios (from 1 to 30) of SPIONs/PEI (assuming PEI was completely combined with SPIONs) with a fixed amount of plasmid DNA encoding VR1020-MSP1$_{19}$ gene (1 µg) for each well plate and incubated in PBS buffer (pH 7.4) for 30 min at room temperature. In addition, SPIONs to PEI mass ratio was fixed to R=10.The SPIONs-PEI-DNA complexes were then added into 6 well plates with a neodymium-iron-boron magnet positioned under each well for about 2 hours. DNA-PEI alone without SPIONs at N/P ratio of (1, 5, and 10) and Lipofectamine 2000-DNA complex were also added in the wells. Western blotting was used in the current work to identify a specific antigen with a specific antibody within a mixture of abundant proteins that are fractionated by using SDS-PAGE and immobilized on the PolyScreen polyvinylidene difluoride (PVDf) membrane.

Results

The SEM and TEM images (Fig.1) show the primary particles with an average diameter of around 10 nm ±3nm, while the measured size from Malvern Zetasizer ZS indicated that the aggregate size of the nanoparticles in suspension was predominantly around 85± 5nm. DLS technique measures the time-dependent fluctuations in the intensity of scattered light from a suspension of particles undergoing random, Brownian motion related to particle size. A possible cause of the larger size measured in suspension, even in the absence of any external magnetic field, was the magnetostatic (magnetic dipole reaction) interactions between particles in aqueous media that may cause aggregation [5].

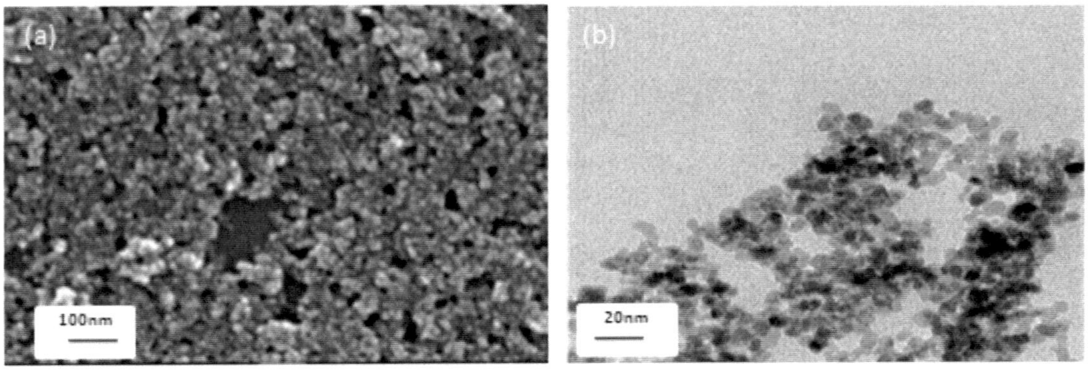

Fig. 1 (A) SEM and (B) TEM images of the synthesized SPIONs

The magnetic measurements of these particles are shown in Fig. 2. Both the uncoated particles and the PEI coated particles had relatively high magnetization and a negligible (0.01emu/g) remanance, indicating that the nanoparticles exhibited superparamagnetic behavior. For iron oxide coated with PEI polymer, the magnetization was not significantly affected by the coating.

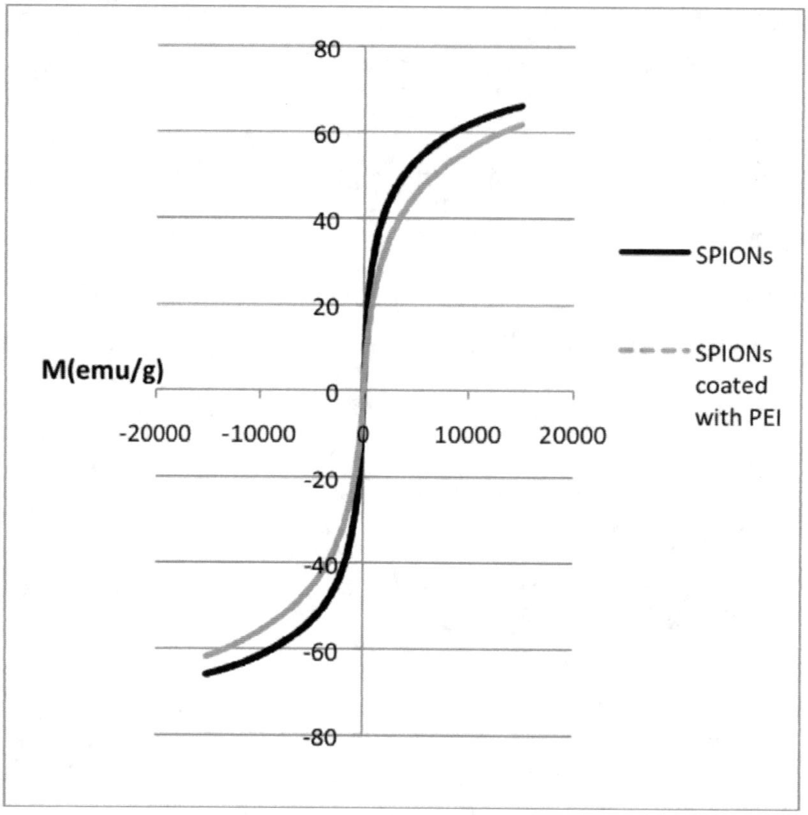

Fig. 2 VSM result of SPIONs and SPIONs/PEI. X and Y axes indicate the applied magnetic field (Oe) and magnetization (emu/g), respectively.

Fig. 3 shows the x-ray diffraction patterns of the sample (naked SPIONs) which is quite identical to standard pattern of pure magnetite JCPDS (Card No. 01-072-6170). There is a low broadening of the diffraction peaks indicating a decrease in the crystallization of magnetite grains. This may be due to the preparation of magnetite nanoparticles in the presence of trisodium citrate that could significantly diminish the crystal growth of the magnetite grains. Another possible cause was the generation of bubbles due to high stirring during the synthesis process, causing oxidation to some parts of the magnetite nanoparticles.

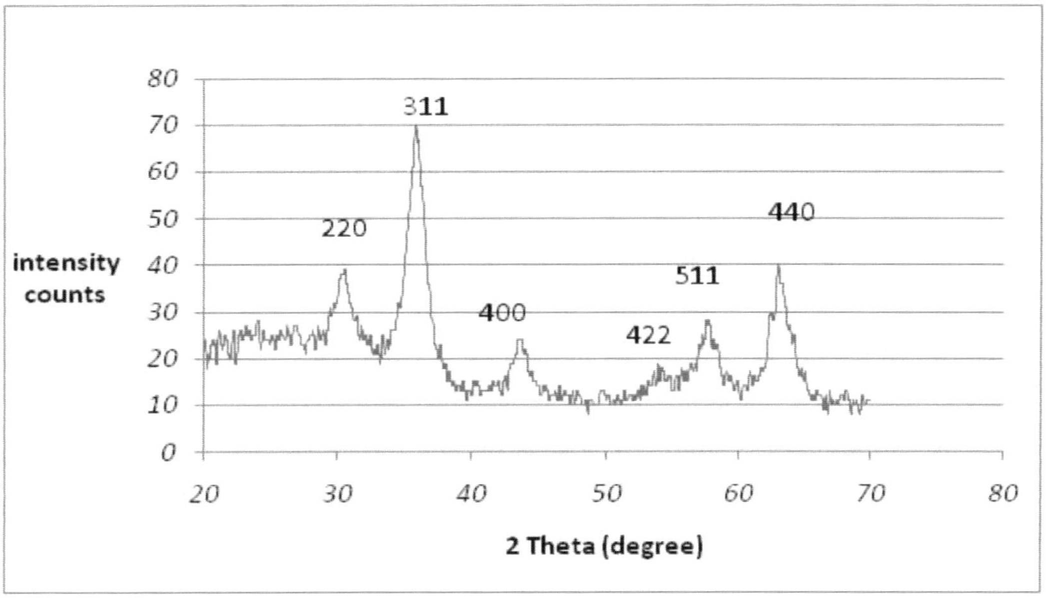

Fig. 3 X-ray diffraction pattern of Fe_3O_4 nanoparticles prepared by co-precipitation method

The stability of the DNA binding was tested with agarose gel electrophoresis. Fig. 4 shows that the DNA at different N/P ratio created a strong complex with PEI, so they could not migrate in all the ratios. PEI has a higher affinity for the DNA than the fluorescent dye used in gel electrophoresis. This implies that the dye was prevented from binding to the plasmid backbone upon the addition PEI.

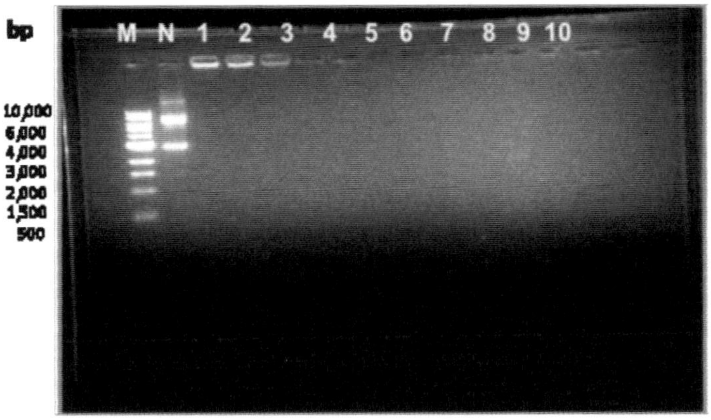

Fig. 4 The stability of SPIONs-PEI-DNA complex by agarose gel electrophoresis at PEI: SPIONs mass ratio equal to 10.
Lane M: DNA marker;
Lane N: naked plasmid $MSP1_{19}$
Lane 1, 2, 3, 4, 5, 6, 7, 8, 9 and 10: N/P ratio of 0.5, 1, 2, 5, 7, 10, 15, 20, 25, and 30, respectively.

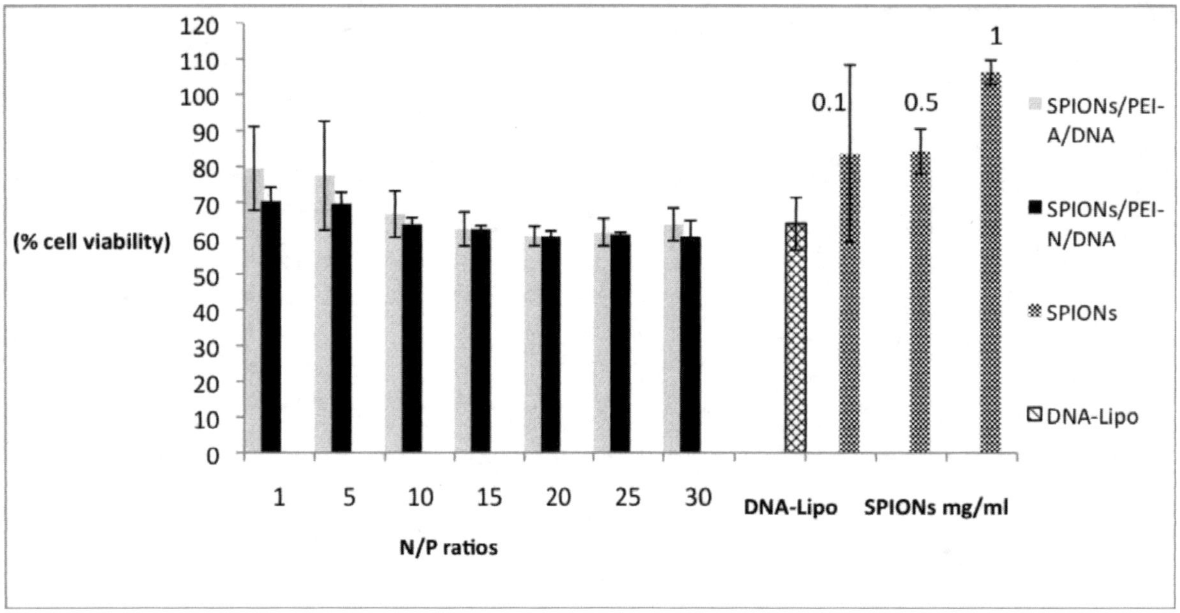

Fig. 5 ***In vitro*** **cytotoxicity of different complexes evaluated in COS7 cells by comparing the percentage of viable cells.**

Fig. 5 shows the influence of DNA-PEI-SPIONs on the cells viability after 24 h exposure to COS7 cell in RPMI medium. More than 60% of the cells remained viable after transfection with DNA-PEI-SPIONs at different N/P ratio under pH 4 .0 and7.0 respectively. This indicates that the pH of complexation had no significant effects on cell viability. It is clear from Fig. 5 that cell viability was slightly less than 80% at an N/P=1 due to the low concentration of PEI that reduced its cell toxicity, however cells viability decreased steadily to around 60% with increasing N/P ratios. Increasing PEI concentration as a result of increasing N/P ratio induced a decrease in cell viability, although there was no significant different in the cell viability between different N/P ratio especially at >10. Lipofectamine 2000 (liposome) is the most effective nonviral reagent examined that yielded consistently high transfection rates accompanied with slightly high toxicity. The cytotoxicity of Lipofectamine is due to 4 protonatable amines on its head group at physiological pH (Dokka et al., 2000). In this study, there was no obvious difference in cell toxicity between Lipofectamine 2000 and SPIONs-PEI-DNA polyplexes especially at high N/P ratios >15. The effect of the naked magnetite nanoparticles in the solution on the cell proliferation was assessed by using different concentration of magnetic solution. Notably more than 83% of cells still remained alive when they were incubated with magnetite solution at different concentration for 24h even with increasing of concentration to 1 mg/ml the cell viability was 100%.

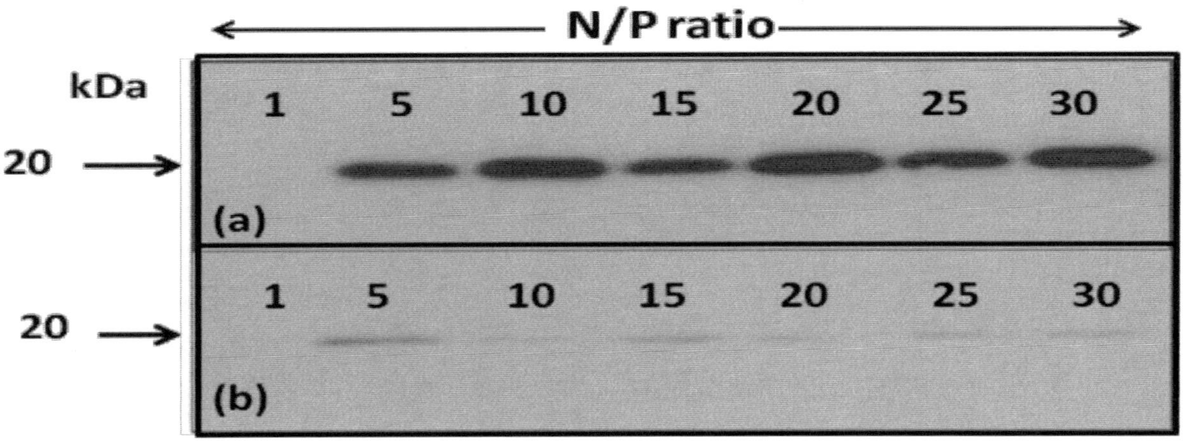

Fig. 6 Western blotting detection of PyMSP1$_{19}$. Lane 1, 5, 10, 15, 20, 25 and 30 corresponding to different N/P ratios of SPIONs-PEI-DNA polyplexes with SPIONs-PEI: (a) with magnet; (b) without magnet.

Based on the western blotting result on the experiment *in vitro,* it is clear from Fig. 6 (a) that the quantity of DNA delivered to the cells at different N/P ratios under magnetic field was 4-6 orders higher. This was due to the drawing of SPIONs/PEI/DNA polyplexes on the surface of the cells via magnetic force, leading to an increase in their cellular uptake. Without the application of magnetic field (Fig. 6b), the gene transfection efficiency using SPIONs/PEI/DNA polyplexes at the same ratios was not enhanced. Moreover, gene expression of the polyplexes was better than the polymer alone with different N/P ratios (data not shown).

Conclusion

Despite the fact that a lot of studies have been done in the field of gene delivery system over decades, much work still needs to be done in the field of DNA delivery for vaccination or therapeutic purpose. In this study, a new magnetofection method for delivery malaria DNA vaccine encoded merozoite surface protein MSP1$_{19}$ has been used *in vitro*. The particles showed a narrow size distribution with little aggregation in the aqueous media and DNA binding functionality rendering them desirable for gene delivery system. Cellular uptake of DNA complex with PEI coating of SPIONs increased dramatically by application of an external magnetic field during the gene transfection process. The high transfection potential of these nanoparticles was demonstrated with magnetofection technique, subsequently leading to high expression of the MSP1$_{19}$ protein *in vitro*. *In vivo* gene delivery experiments will be conducted to investigate the efficiency of magnetofection principle by using these superparamagnetic magnetite nanoparticles to incite high immune responses following the polyplexes vaccination in mice.

References

1. Pimentel, D., et al., *Ecology of increasing diseases: Population growth and environmental degradation.* Human Ecology, 2007. **35**(6): p. 653-668.
2. Snow, R.W., et al., *The global distribution of clinical episodes of Plasmodium falciparum malaria.* Nature, 2005. **434**(7030): p. 214-217.
3. Boussif, O., et al., *A VERSATILE VECTOR FOR GENE AND OLIGONUCLEOTIDE TRANSFER INTO CELLS IN CULTURE AND IN-VIVO - POLYETHYLENIMINE.* Proceedings of the National Academy of Sciences of the United States of America, 1995. **92**(16): p. 7297-7301.
4. Behr, J.-P., *The Proton Sponge:a Trick to Enter Cells theViruses Did Not Exploit.* Chimia, 1997. **51**: p. 34-36.
5. Mahmoudi, M., et al., *Cell toxicity of superparamagnetic iron oxide nanoparticles.* Journal of Colloid and Interface Science, 2009. **336**(2): p. 510-518.

(656b) Targeted Virus Nanoparticles for Localized Chemotherapy of Breast Cancer

Fang Wei[1], Kellie I. McConnell[1], Tse-Kuan Yu[2] and Junghae Suh[1], (1)Bioengineering, Rice University, Houston, TX, (2)Radiation Oncology, The University of Texas M. D. Anderson Cancer Center, Houston, TX

A critical difficulty in designing nanotherapeutics is in developing highly targeted platforms that can reach target sites with high efficiency. Additionally, due to the high degree of heterogeneity in breast tumors within a patient as well as between patients, there is currently no one biomarker available that specifically identifies all breast cancer cells. To overcome these obstacles, we have developed a novel drug delivery system based on adeno-associated virus (AAV) to deliver chemotherapeutic agents specifically to breast cells and limit the exposure of drugs to normal non-breast tissues. AAV is a 25 nm virus that is currently in clinical trials for a variety of gene therapy applications. The AAV capsid is a supramolecular assembly of 60 protein subunits, lending itself well to multivalent conjugation of drug molecules. We have used directed evolution to create virus nanoparticles that can specifically bind breast tissue. Paclitaxel, chemically modified with N-hydroxysuccinimide, was covalently attached to the virus capsid surface. Drug-virus nanoparticles were assayed for degree of conjugation with UV-vis and MALDI-MS, proper capsid assembly with ELISA and TEM, genomic packaging with Q-PCR, and cytotoxicity with cell viability assays. This work will offer superior targeting of breast tissues and provide greater drug payload. It has the potential of significantly decreasing the morbidity associated with systemic chemotherapy.

See more of this Session: Nanomedicine and Drug Delivery V
See more of this Group/Topical: Topical F: Nanotechnology in Medicine

(656c) Magnetic Hydrogel Nanocomposites for Synergistic Chemotherapy and Hyperthermia-Based Treatment of Cancer

Samantha A. Meenach, J. Zach Hilt and Kimberly W. Anderson, Chemical and Materials Engineering, University of Kentucky, Lexington, KY

Hyperthermia, the heating of tissue to 41 to 45°C, has been shown to improve the efficacy of cancer therapy when used in conjunction with irradiation and/or chemotherapy. This can be done through the utilization of biomaterials which can be delivered at tumor sites and remotely heated from outside the body while simultaneously controlling the delivery of therapeutic drugs. One such potential class of materials is magnetic hydrogel nanocomposites which are composed of biocompatible polymeric systems and magnetic nanoparticulates capable of heating upon exposure to an electromagnetic field. In this research, hydrogel nanocomposites have been developed that can be implanted or injected within a tumor to control the delivery of both heat and chemotherapeutic agents. The nanocomposites studied for this project include a stealth, poly(ethylene glycol) (PEG)-based system and a poly(β-amino ester) (PBAE)-based biodegradable system. Iron oxide nanoparticles were physically entrapped in these materials during polymerization to provide the remote heating mechanism of these systems. The capability of the hydrogels to be heated in an alternating magnetic field at various temperatures depending on the strength of the field was demonstrated. Swelling analysis was done and showed temperature-responsive behavior of PEG gels in their swollen state. These gels were determined to be non-toxic to murine fibroblasts indicating their potential biocompatibility. M059K glioblastoma cells were heated to thermoablative temperatures (above 55°C) via gels exposed to the AMF, and the resultant cell death was observed. PBAE degradable gels have been evaluated that will have the ability to heat via the iron oxide nanoparticles and have controlled release of chemotherapeutic drugs. The degradation profile of these gels can be tailored by the ratio of macromers used, and again, remote heating has been successfully demonstrated. In summary, these systems have the ability to be remotely heated and to deliver drugs to specific sites, which make them viable systems for treating deep-seated or reoccurring tumors.

See more of this Session: Nanomedicine and Drug Delivery V
See more of this Group/Topical: Topical F: Nanotechnology in Medicine

METHOD TO SYNTHESISE, OPTIMISE AND CHARACTERISE SMART MULTI-FUNCTIONAL MAGNETIC NANOPARTICLES FOR CANCER TARGETTING

Bhushan Shinde, Dr. Rachna Rastogi, Dr Veena Koul, Dr. Ashok Bhaskarwar, Indian Institute of Technology, Delhi, India

Introduction

Cancer was once considered an incurable disease, but today most patients diagnosed with early stage disease can have chances for survival. Advances in cancer diagnostics and therapeutics over the last few decades are largely responsible for this dramatic improvement. This is shown by the five-year survival rate for all cancers diagnosed between 1996 and 2002, which is 66%, up from 51% in 1975-1977 in US. In India, there used to be about 850 000 new cancer cases diagnosed every year resulting in about 580 000 cancer related deaths every year. Over the years this number is decreasing with better care but cancer still remains amongst the leading cause of death in India and worldwide.

Over the past decade, there has been an increasing interest in using nanotechnology for cancer therapy. The development of smart targeted nanoparticles (NPs), polymer-drug conjugates, hydrogel loaded drugs etc. that can deliver drugs at a sustained rate directly to cancer cells may provide better efficacy and lower toxicity for treating primary and advanced metastatic tumors.

This work deals with a method to synthesize, optimize and characterize a novel drug delivery system. This system encapsulates a hydrophobic drug within thermally sensitive polymer nanoparticles with magnetic sensitivity. This multiple responsive system is effective in enhancing the efficacy of the hydrophobic drug and in controlling its release to the tumor regions.

Smart materials responsive to multiple environmental stimuli are of interest to biotechnology because of possible applications such as delivery carriers, separation platforms and environment sensors. Since body temperature is nearly constant, a small temperature excursion about it provides an environmental stimulus to be exploited. Temperature-responsive soft materials used in conjunction with localized heating (e.g., via hyperthermia) are therefore prime candidates for biomedical applications. Other stimuli such as pH, glucose, stress or strain, and electromagnetic fields can be combined with thermal stimulus to create a multi-stimuli-responsive system. Here we focus on magnetic stimulus which can be applied remotely.

One possible application of magnetically and thermally responsive smart nanomaterials is illustrated in Fig. 1 that pertains to remotely controlled drug delivery. A soft material/hard material hybrid in which a magnetic metal or oxide is embedded in a temperature-responsive polymer matrix can combine the thermal sensitivity with magnetic signatures. Importantly, nanosizing such construct brings about new desirable features of extremely fast thermal response time, small magnetic hysteresis and enhanced magnetic susceptibility.

The hard magnetic material of choice is iron oxide, which is relatively safe for biomedical applications and can be readily synthesized in a form of small particles to be embedded into the soft material. The temperature responsive polymer matrices has been chosen to be of a hydrogel matetrial of N-isopropylacrylamide (NIPAAM).

The main objective of the proposed work was to synthesize, optimize and characterize a novel multi-responsive drug delivery system and study the in vitro release kinetics of a hydrophobic drug loaded onto the system.

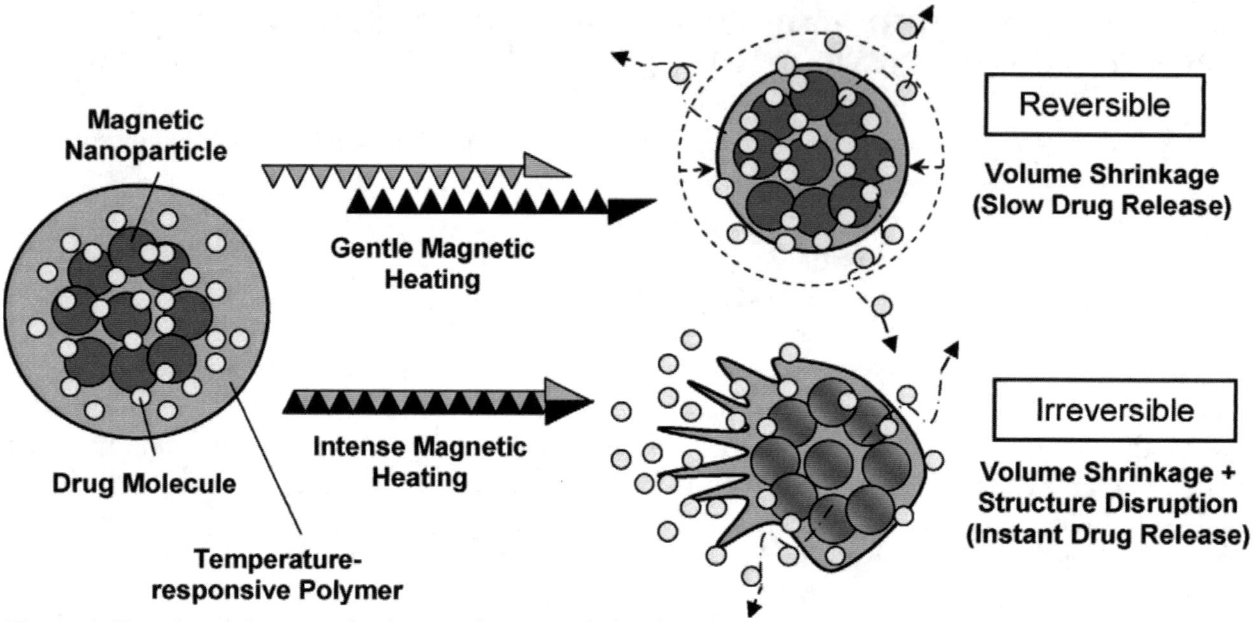

Figure 1. Two drug release mechanisms under magnetic heating. Gentle magnetic heating causes temperature-responsive polymer to shrink, squeezing drug out from the nanoparticle. Intense magnetic heating additionally ruptures the nanoparticle, triggering a burst-like drug release.

Materials and Methods

Materials

The chemicals listed below were purchased from Sigma- Aldrich, USA and were procured from local vendors. These were used either as-received or further purified as indicated below. The chemicals used as-received are: iron (III) chloride ($FeCl_3$), iron (II) chloride ($FeCl_2$), potassium hydroxide (KOH), ammonium persulphate (APS), tetramethylethylenediamine (TEMED), N,N'-methylene-bis-acrylamide (BIS). N-isopropylacrylamide (NIPAAm) was recrystalised in hexane. Deionised water purged with nitrogen gas was used in all steps involved.

Synthesis of iron oxide nanoparticles

The following procedure was used for synthesis of iron oxide nanoparticles. 0.42 g $FeCl_2$ and 1.09 g $FeCl_3$ ($Fe^{3+}/Fe^{2+} = 2.59$) was dissolved in 1N HCl. 10 ml of 1N KOH was added drop wise to this solution (40 ml) of under N_2 atmosphere leading to a black precipitate. The stirring rate was 400 RPM.

Synthesis of blank thermosensitive polymer nanoparticles

The following procedure was used for synthesis of the PNIPAAM nanoparticles (NPs) separately. The solution of 0.25g NIPAAm (0.002 moles), in water with 1% N, N dimethylene bis acrylamide (BIS; cross linker) and 10% sodium lauryl sulphate (SLS) was prepared in 50 ml deionised water. The solutions were purged with nitrogen for 30 minutes to create an inert atmosphere. Finally 1% by moles ratio ammonium persulphate (APS) and tetramethylethylenediamine (TEMED) each (initiator) were added and the reaction was carried for 4 hours at 70 ^{o}C. The stirring rate was 400 RPM.

Magnetic nanoparticles encapsulated in thermosensitive polymer naoparticles

To synthesise magnetic nanoparticles the following procedure was adopted: To a solution of .25g NIPAAm (0.002 moles) and 0.0792g PEGMA (0.0002 moles) in 40 ml water with 1% by moles ratio Bis acrylamide (cross linker) and 1% by moles ratio SLS was made in 40 ml DI water. 0.42 g $FeCl_2$ and 1.09 g $FeCl_3$ ($Fe^{3+}/Fe^{2+} = 2.59$) and different ratios accordingly for 1:25 and 1:50 iron: monomer ratios were added by mechanical stirring. 10 ml of 1N KOH was added drop wise to this solution (40 ml) of under N_2 atmosphere leading to a black precipitate. Then 1% by moles ratio APS and TEMED each (initiator) were added and the reaction was carried for 4 hours at 70 °C. The stirring rate was 400 RPM.

Synthesis of nanoparticles

A fixed co-polymer mole ratio of 1:10 (NIPAAM: PEGMA) was selected. Nanoparticles of NIPAAm-PEGMA (NP) and their magnetic counterparts (MNPs) were prepared as described in the experimental section.

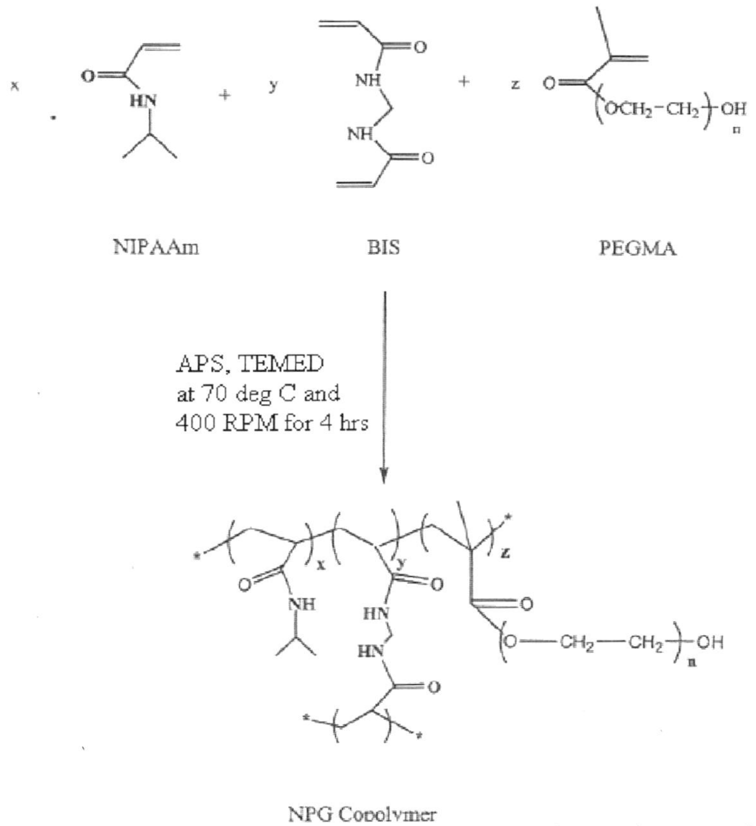

Fig 2 Reaction scheme for the copolymer formation reaction

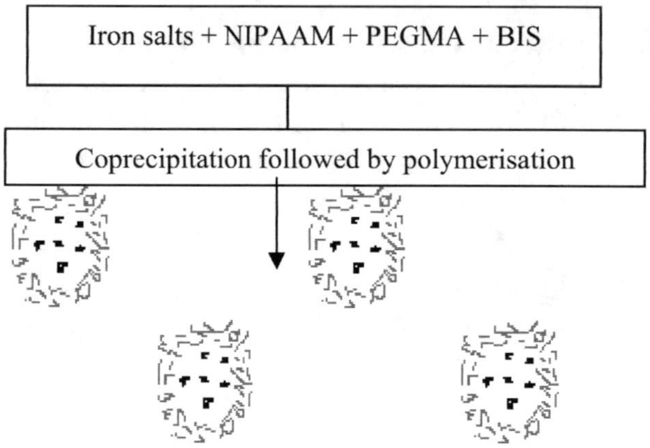

Fig 3 Schematic diagram for the formation of magnetic nanoparticles encapsulated by thermosensitive polymer nanoparticles.

The prepared samples were then characterized for size, zeta-potential, morphology by TEM and AFM.

Characterisation

Particle Size and zeta potential

a) Iron oxide (Fe_2O_3) nanoparticles with oleic acid coating (8 nm)

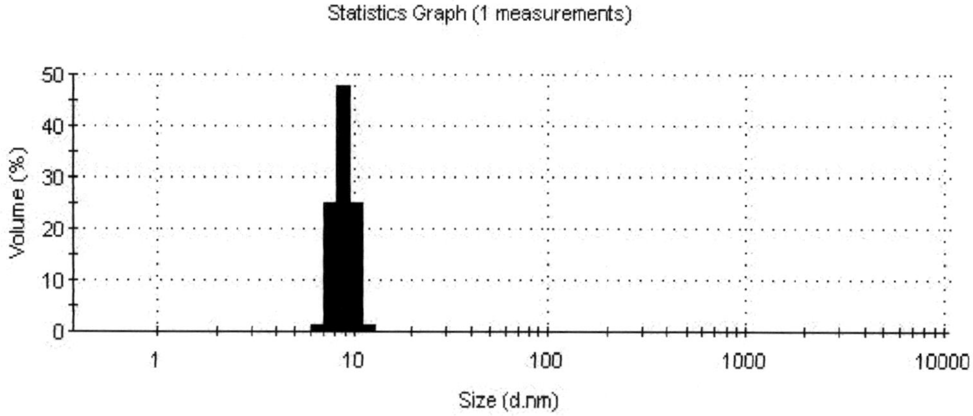

The iron oxide nanoparticles were successfully synthesised by co-precipitation technique. Then they were washed with water to bring the pH below 9. Further they were centrifuged and dried in oven at 40 ^{0}C for 24 hrs. The DLS measurements were done by making a suspension in dichloromethane and the average size was found to be 8 nm PDI 0.32.

b) Blank Thermosensitive polymer nanoparticles (140 nm)

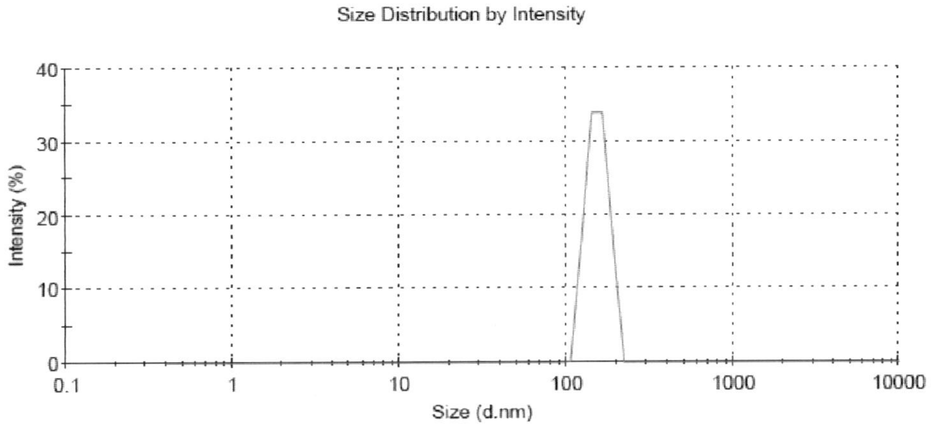

The blank thermosensitive polymer nanoparticles were synthesied with the help of Bis-acrylamide (cross-linker) and Ammonium persulphate and tetramethylethylenediamine (initiator) at 70 ^{0}C for 4 hrs as mentioned in the Methods section. The DLS measurements gave the average size as 140 nm PDI 0.24.

c) Thermosensitive nanoparticles encapsulating magnetic nanoparticles (Effect of iron: polymer ratio)

Iron: Polymer (Mole ratio)	Z_{avg} (nm)	Poly Dispersity Index (PDI)	Peak (nm)
1:100	312	0.366	288.9
1:50	289	0.131	298.7
1:25	222	0.137	241.8

The samples (1:25 and 1:50) were selected for further characterization due to their lower sizes and polydispersities.

d) Zeta potential of 1:50, 1:25 and blank thermosensitive polymer nanoparticles

The zeta potential changed from negative for blank thermosensitive polymer particles to positive when iron concentration was increased in the nanoparticles.

Transmission Electron Microscopy

TEM of Iron oxide nanoparticles coated by thermosensitive polymer nanoparticles

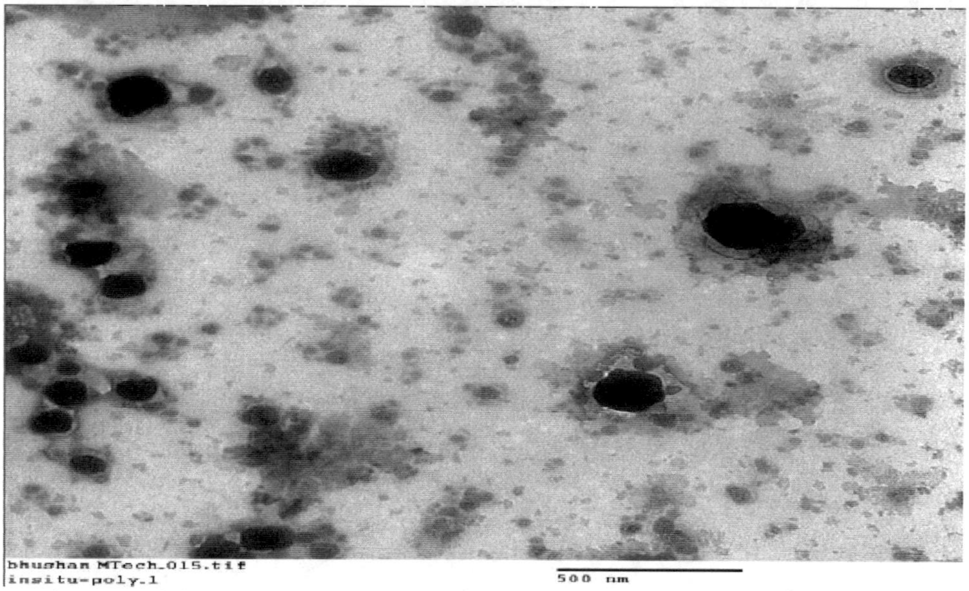

The TEM image of thermosensitive polymer coated iron oxide nanoparticles is shown above. There was no aggregation as in the case of uncoated iron oxide nanoparticles and the polymer coating can be clearly seen. The size of the particles is in the range of 150-250 nm.

Atomic Force Microscopy

1:25 (Fe: Monomer)

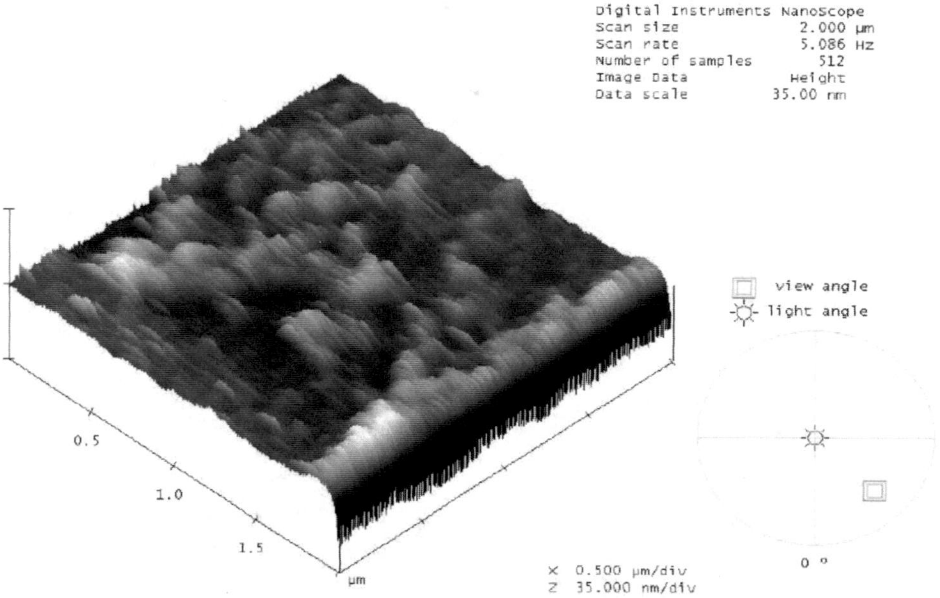

As can be seen, the particles are visible in 1:25 image with the contact mode AFM for dry sample.

pH Stability

a) pH 5.8

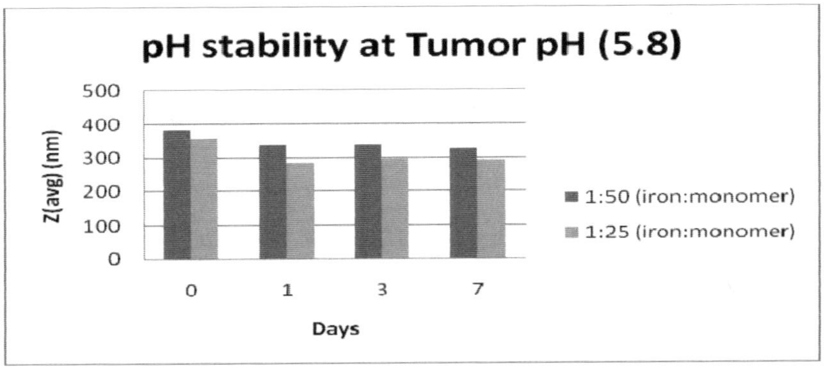

b) pH 7.4

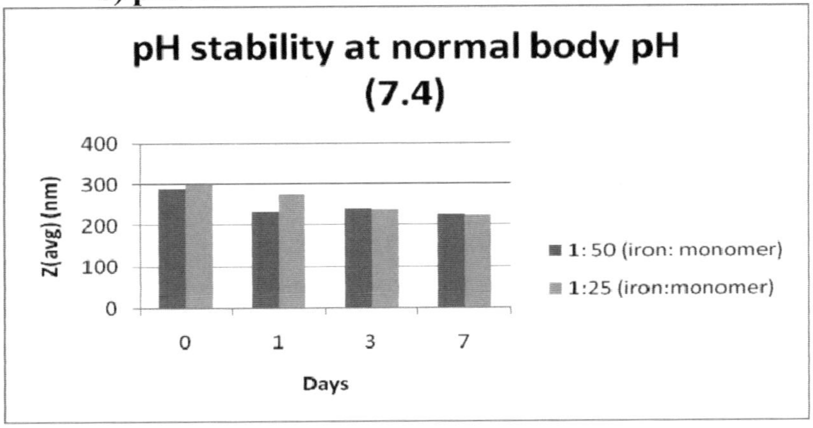

As can be seen, both the nanoparticles were stable in size at the tumor pH and the body pH over the period of testing of 7 days.

FTIR Spectroscopy

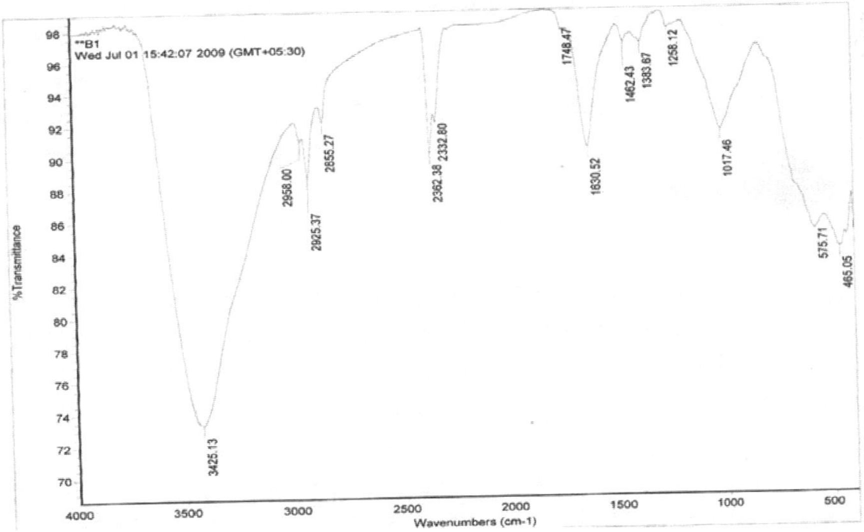

The FTIR spectrum of MNPs is displayed in Figure... The Amide I band of NIPAAm and BIS appeared at 1630.52 cm^{-1}. A peak at 1748.47 cm^{-1} corresponding to the C=O vibrations of NIPAAM, BIS and PEGMA was noted. The presence of PEGMA can be deduced by the broad peak at 1017.46 cm^{-1} denoting the C-O-C linkages in the structure. The methyl and methylene stretching vibrations of the polymer were observed at 2958, 2925.37 and 2855.27 cm^{-1}. The FTIR spectrum showed the presence of significant quantities of iron oxide. The absorption peak at 583cm^{-1} was observed, which can be ascribed to Fe_3O_4. Depending on Fe (II) content, this pattern is consistent with the magnetite (Fe_3O_4) spectrum (band 570 cm−1) or maghemite (γ-Fe_2O_3) spectrum (broad band 520–610 cm^{-1}).

Magnetisation curve for the 1:25 ratio sample at 40 ^{0}C

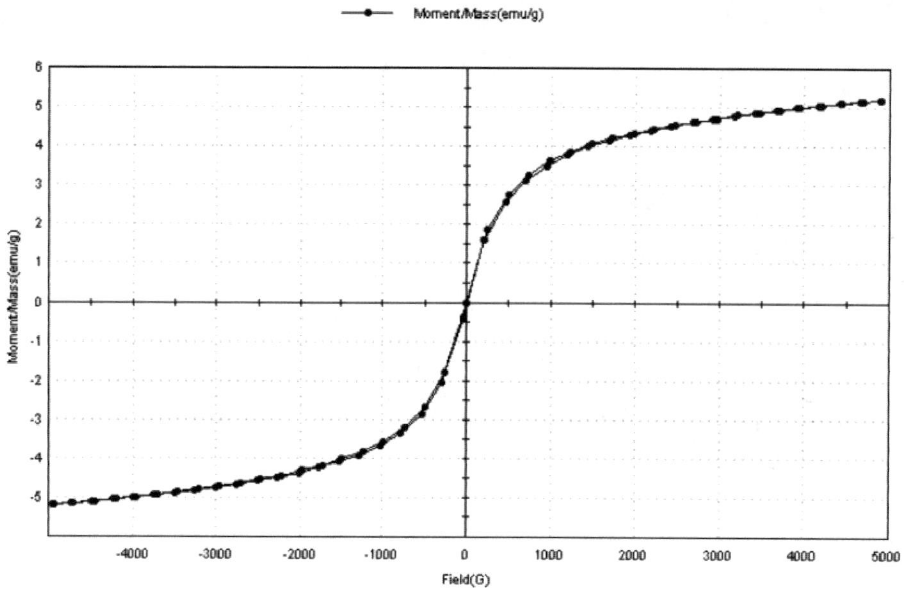

The relative magnetization curve as a function of magnetic field for the 1:25 coated polymer at 40 ^{0}C is shown above. From the figure, no hysteresis curve was observed, which indicates the characteristic superparamagnetic behavior of the particles. The saturation magnetization value of the magnetic nanoparticles was found to be 5.2 electromagnetic units per gram (emu/g). In addition the magnetization decreases from the plateau value and reaches zero when the magnetic field is removed.

Drug release result and analysis

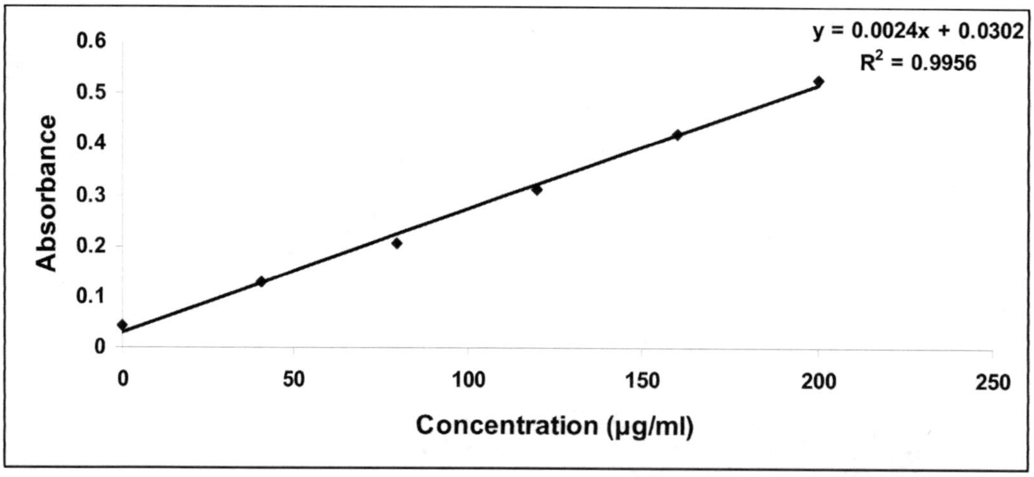

Fig 3 Standard plot of curcumin at 429 nm in ethanol:water (3:2) system

Standard plot of curcumin was taken at 429 nm in ethanol:water system. The correlation coefficient (R^2) was found to be 0.9956. This plot was further used to determine the concentrations of the samples collected from the drug release study.

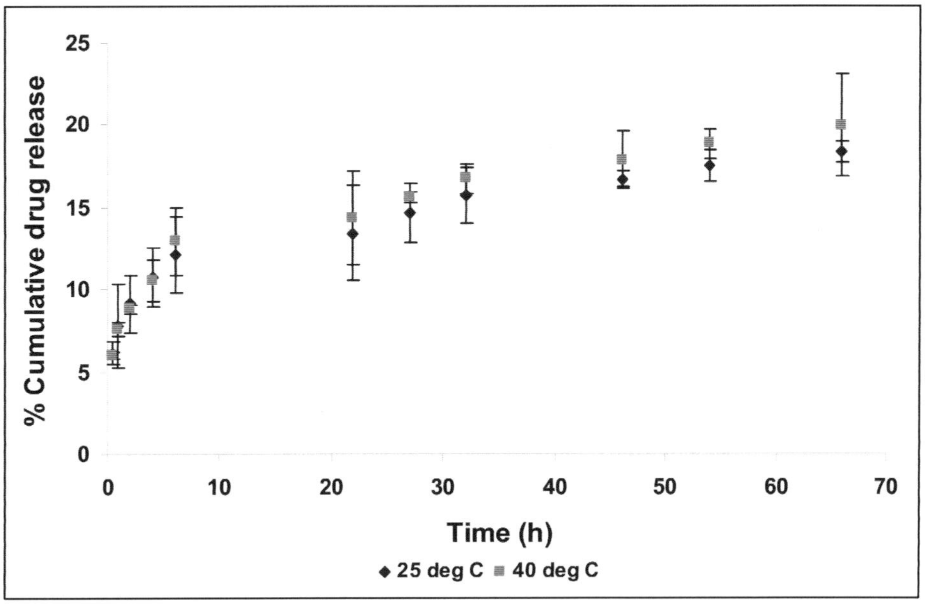

Fig 4 Extent of drug released from MNPs (1:25) in 4% BSA solution at 25 and 40 °C.

The loading efficiency of curcumin for 1:25 MNPs was found to be 0.466 mg/mg nanoparticles, that is, 46.66%. Drug release studies were carried out in 4% BSA solution at 25 and 40 °C. The amount of curcumin released was determined from the standard plot as described above. The release was found to be controlled at both the temperatures and no burst release was noted. The total percentage of drug released was 18.3 and 20.9 at 25 and 40 °C respectively after a period of 66 h. The release of curcumin was approximately 1.4 times higher at 40 °C.

Extraction of the remaining drug after in vitro studies showed nearly 60.45 and 57.1 % of curcumin for the two temperatures. The loss of drug was due to its binding to albumin used for the release studies.

Investigating the model of drug release

Model	R^2 value
First order model	0.876
Zero order model	0.781
Higuchi Matrix model	0.847
Peppas-Korsmeyer Equation	0.845
Hixson-Crowell Equation	0.872

Fig 5 Model of drug release

From figure 5 we find that the first order model and the Hixson-Crowell model have the highest R^2 Value. But the Hixson-Crowell Equation is not applicable in this case as it is for dosage form in the form of tablets and not for nanoparticles. Hence the first order model is found to be followed. The reasoning behind this choice is that in the first order model the release is dependent on the concentration of the diffusing substance. In this case, curcumin which is hydrophobic interacts with the hydrophobic NIPAAM of the shell and thus the release is dependent upon curcumin's interaction and thus its concentration.

Conclusion

A novel multi-responsive drug delivery system was successfully synthesized by chemical methods. The system was optimized with respect to the iron: monomer ratio for optimum size, stability and magnetic property. The characterization of the nanoparticle system was done by DLS, TEM, AFM, FTIR, TGA, pH stability and temperature stability. The in vitro release of a hydrophobic drug (curcumin) loaded onto the system was studied. Finally, the drug release kinetics was studied based on the models proposed in the literature and the system was found to follow a first order drug release property.

References

Gu FX, Karnik R, Wang AZ, Alexis F, Langer RS, and Farokhzad OC (2007), "Targeted nanoparticles for cancer therapy", *Nano Today,* 2, 3, 14-23

Tartaj P, Morales MP, Gonzalez-Carreno T, Veintemillas-Verdaguer S, Serna CJ, (2005), "Advances in magnetic nanoparticles for biotechnology applications", *Journal of Magnetism and Magnetic Materials* 290–29, 28–34

Xiao-Min L, Anna CS, Samiab, (2006), "Synthesis, assembly and physical properties of magnetic nanoparticles", *Journal of Magnetism and Magnetic Materials* 305, 100–109

Giria J, Guha Thakurtaa S, Bellare J, Nigam AK, Bahadur D, (2005), "Preparation and characterization of phospholipid stabilized uniform sized magnetite nanoparticles", *Journal of Magnetism and Magnetic Materials* 293, 62–68

Marsha AM, Brem H, Langer R, (2003), "Advancing the field of drug delivery: Taking aim at cancer" *Cancer Cell·* 4, 337-341

Gaucher G, Dufresne MH, Leroux JC, (2005), "Block copolymer micelles: preparation, characterization and application in drug delivery", *Journal of Controlled Release* 109, 169–188

Schmaljohann D, (2006), "Thermo- and pH-responsive polymers in drug delivery", *Advanced Drug Delivery Reviews*, 58, 1655–1670

Okassa LN, Marchais H, Chourpa I, (2007), “Optimization of iron oxide nanoparticles encapsulation within poly(D,L-lactide-co-glycolide) sub-micron particles”, *European Journal of Pharmaceutics and Biopharmaceutics* 67, 31–38

Nishiyama N, Kataoka K, (2006), “Current state, achievements, and future prospects of polymeric micelles as nanocarriers for drug and gene delivery”, *Pharmacology & Therapeutics,* 112, 630–648

Qiu Q, Somasundaran P, (2002), “Complexation between Poly(maleic acid/octyl vinyl ether) and Poly(vinyl caprolactam) in Aqueous Solution and at the Alumina/Water Interface”, *Journal of Colloid and Interface Science* 253, 231–237

Soga O, Nostrun CF, (2005), “Thermosensitive and biodegradable polymeric micelles for paclitaxel delivery”, *Journal of Controlled Release*, 103, 341–353

(656e) Nanopolyplexes - Responsive Multipurpose Delivery Vehicles

Hitesh G. Bagaria, Dept of Chemical and Biomolecular Engineering, Rice University, Houston, TX and Michael S. Wong, Chemical and Biomolecular Engineering and Chemistry, Rice University, Houston, TX

Our body presents several mechanisms that prevent delivery systems from reaching their target. Overcoming these hurdles requires the development of intelligent delivery vehicles that could respond to the each of the mechanisms. We envision a novel self-assembly based technique for synthesizing nano-polyplex delivery systems that could help towards this goal. This technique involves assembly of polycationic (negatively-charged) polymers by multivalent anions (positively-charged) to form polymer-salt complexes. By this method, size and composition of the delivery system is altered easily and, cargos as varied as siRNA, flurochromes and drugs can be encapsulated. These nano-polyplexes can be fabricated to be responsive to specific stimulus, which enables them to deliver the cargo at desired sites. Here we present two examples of nano-polyplexes for: (i) delivery of siRNA and (ii) development of protease-responsive nanoprobes for imaging.

See more of this Session: Nanomedicine and Drug Delivery V
See more of this Group/Topical: Topical F: Nanotechnology in Medicine

(656f) Dendrimer-Based Nanodevices for the Treatment of Neuroinflammation in Cerebral Palsy

Rangaramanujam M. Kannan[1], Raghavendra Navath[2], Hui Dai[3], Bindu Balakrishnan[4], Roberto Romero[5] and Sujatha Kannan[3], (1)Chemical Engineering, NICHD PRB, Wayne State University, Detroit, MI, (2)Chemical Engineering/NICHD PRB, Wayne State University, Detroit, MI, (3)Pediatrics, NICHD PRB, Wayne State University, Detroit, MI, (4)Wayne State University, Detroit, MI, (5)Eunice Kennedy Shriver National Institute of Child Health and Human Development, NICHD, NIH, DHHS, Detroit, MI

Maternal intrauterine inflammation resulting in microglial activation has been implicated in the development of periventricular leukomalacia and cerebral palsy. N-acetyl cysteine (NAC) is a drug that is currently being explored for the treatment of neuroinflammation in neonatal and perinatal applications. However, plasma binding of NAC significantly reduces the bioavailability requiring very high doses (100-300 mg/kg in animal models). We have developed neutral PAMAM dendrimer-based nanodevices where a disulfide linker is used to link the drug to the dendrimer. This enables tailored intracellular release of the drug in a manner sensitive to the glutathione levels (low in blood circulation, high inside the cells). When this is combined with the ability of dendrimers to selectively localize in activated microglial cells, significant improvements in vivo performance is achieved. The nanodevices are evaluated extensively in a rabbit model of cerebral palsy. The biodistribution and efficacy of intravenously administered dendrimer-drug conjugates are compared with those of the free drug using a combination of tools. The biodistribution is studied using microPET imaging and tissue confocal microscopy. The efficacy is evaluated using a combination of neurobehavioral analysis, assessment of brain tissue level inflammatory cytokine analysis. Our studies show that dendrimer-drug conjugates are 10-100 times more efficacious that free drug, suggesting that these conjugates (~ 18000 Da) are able to cross the blood brain barrier and deliver the drug significantly better than free drug. The dramatic improvements in motor deficits of dendrimer-drug treated animals suggest that targeted and timely treatment of neuroinflammation can lead to significant improvements in therapeutic outcomes in this disease with no known cure. Detailed mechanistic and efficacy studies will be presented.

(1) 'Drug release mechanisms and kinetics from dendrimer-drug conjugates with glutathione sensitive linkers', Emre, YK, R. Navath, B. Wang, R. Romero, S. Kannan, RM Kannan*, Biomaterials, 30, 2112-2121 (2009) (2) 'Dendrimer-drug conjugates for tailored intracellular drug release based on glutathione levels', Navath, R; E. Turkoglu; B. Wang; S. Kannan;R. Romero; R.M. Kannan*, Bioconjugate Chemistry, 19, 2446-2455 (2008) (3) Intrinsic targeting of neuroinflammation by polyamidoamine dendrimers in a rabbit model of cerebral palsy' H.Dai, R.Navath, B.Balakrishnan, B.Raja Guru, M. Mishra, R.Romero, R.M.Kannan*, S.Kannan*, revised and submitted, Future Medicine:Nanomedicine, Apr. (2010) (4) Dendrimer-based therapeutic nanodevices for therapeutic and imaging applications, R. Kannan, S. Kannan, R.Romero, R. Navath, H. Dai, Y. Kurtoglu, B.Wang, A. Menjoge, Provisional patent filed # 61/187263, 5/09, and additional provisional patent filed, #61/319285, 3/10

(656g) Synthetic Platelets for Biomedical Applications

Nishit Doshi[1], Jennifer Orje[2], Zaverio Ruggeri[2] and Samir Mitragotri[1], (1)Chemical Engineering, University of California, Santa Barbara, Santa Barbara, CA, (2)Department of Molecular and Experimental Medicine, The Scripps Research Institute,, CA, La Jolla, CA

Arterial and venous thrombi result in life threatening complications depending on the site of formation or lodgment of embolus. These include cardiovascular diseases such as myocardial infarction and stroke that form the leading cause of deaths in United States. Thrombus formation, initiated by circulating platelets, is an important and initial step in repairing injured vasculature, however, excess clot formation results in pathological conditions such as thrombosis. Platelets perform several important biological functions such as hemostasis, release of growth factors and modulation of inflammatory processes. The unique physical features of platelets play an important role in their biological function. Here, using novel fabrication techniques, we have created synthetic replicates of platelets (sPlats) which mimic the important physical, chemical and biological attributes of natural platelets. sPlats have a size of 2-4 µm, a discoidal shape and a highly flexible membrane similar to natural platelets. sPlats are fabricated by layer-by-layer deposition of positively and negatively charged polyelectrolytes including proteins such as actin, an inherent component of the platelet cell membrane and poly-allyl amine hydrochloride on polymeric particles followed by dissolution of the polymeric core. Similar to their natural counterparts, GP1b coated sPlats show excellent adhesion propensity to VWF coated substrate. GP1b is the receptor on platelets that binds strongly to VWF expressed at the site of vascular injury. The adhesion propensity of sPlats was significantly higher than spheres, discoids and albumin coated sPlats indicating the importance of size, shape, mechanical flexibility and surface chemistry of sPlats. sPlats hold outstanding potential in various biomedical applications. Since sPlats mimic the important properties of natural platelets, they can serve as ideal drug delivery vehicles in the treatment of conditions associated with either a low platelet count or high platelet activity. The highly selective adhesion propensity of sPlats to VWF domain coupled with the ability to release growth factors make them excellent substitutes to natural platelets. Mimicking the structural properties of natural platelets may enable sPlats to exhibit prolonged circulation and high targeting efficiency, the properties highly desired in a drug delivery vehicle. sPlats hold promise in treatment of thrombosis by selectively attaching to the clot site and releasing thrombolytic drugs in a controlled manner. sPlats can also be tracked in real time since they can be encapsulated with iron oxide nanoparticles which show high contrast in magnetic resonance imaging. The fabrication of these novel platelet mimetic particles has opened up new avenues in the fields of drug delivery, diagnostic imaging, treatment of cardiovascular diseases and other platelet associated disorders and artificial blood synthesis.

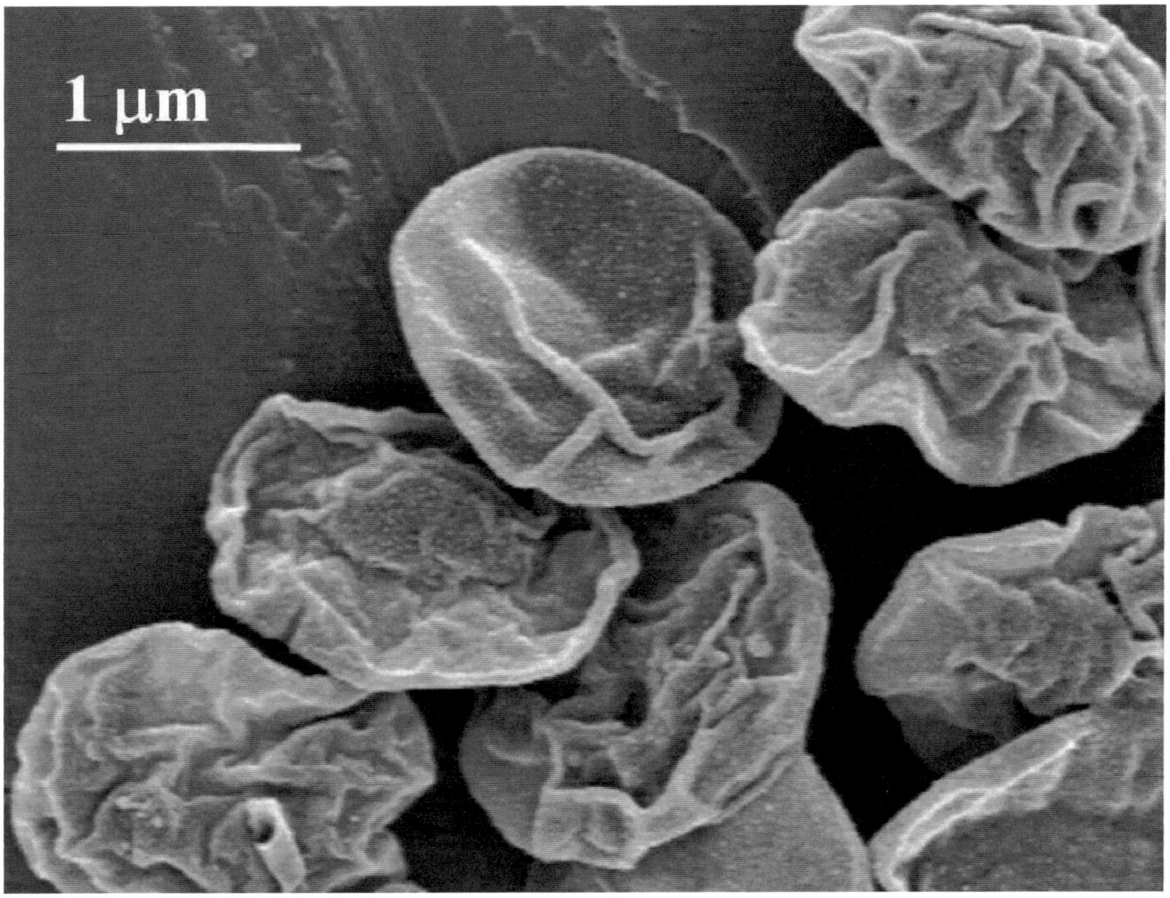

See more of this Session: Nanomedicine and Drug Delivery V
See more of this Group/Topical: Topical F: Nanotechnology in Medicine

(701a) Nanobiosensors

Ajit Sadana, Chemical Engineering Department, University of Mississippi, University, MS

Applications of nanomataerials to nanobiosensors are reviewed. Particular emphasis will be placed on the enhancement of biosensor parameters such as sensitivity, selectivity, economics, response time etc, if any. Some of the examples presented and analyzed include: (a) Multi-function cross-linked Au nanoagregates for optical DNA detection (Li et al., 2009), (b) Nano-bio-chips for high performance multi-plexed protein detection: determination of cancer markers in serum and saliva using quantum dot and bioconjugate labels (Jokers et al., 2009), (c) Nonenzymatic glucose sensor based on renewable electrospun Ni nanopaticle-loaded carbon nanofiber paste electrode (Liu et al., 2009),(d)The development and in-vitro characterization of an intracellular microsensor responsive to reactive oxygen species (Henderson et al,. 2009), (e) Nanotube-based biosensor for the detection of disease-specific autoantibodies in human serum (Drouvalakis et al., 2008), (f) Periodic plasmonic nanostructures as efficient SERS (surface enhanced raman spectra) substrates for biosensing (Lin et al, 2009), (g) DNA hybridization detection using spectral chnages of zinc-selenide nanocrystals (Wang et al., 2009), and (h) Layer-by-layer assembly of multiwall carbon nanotube ultrathin films for biosensing (Mantha et al., 2009). The examples presented together and briefly analyzed provide one with a perspective of the different nanotechnology techniques and processes that have been recently used to develop effective nanobiosensors. The examples selected were at random. With the current emphasis on the detection of biomarkers for different diseases such as MI (myocardial infarction), and authoimmune diseases such as RA (rheumatoid arthritis) and SLE (systemic lupus erythematosus), medical applications of bisensors wil continue to dominate the market, and the application of nanobiosensors to this area is not to be entirely unexpected. However, applications of nanobiosensors to non-medical areas will also continue to increase due to the ease of applications

See more of this Session: Nanomaterials
See more of this Group/Topical: Topical F: Nanotechnology in Medicine

(701b) Use of Naturally-Occurring Halloysite Nanotubes for Enhanced Capture of Cancer Cells From Blood

Andrew D. Hughes and Michael R. King, Biomedical Engineering, Cornell University, Ithaca, NY

The development of individualized treatments for cancer can be facilitated by more efficient methods for separating cancer cells from patient blood in such a way that they remain viable for live cell assays. We have previously shown that immobilized P-selectin protein can be used on the inner surface of a microscale flow system to induce leukemic cells and leukocytes to roll at different velocities and relative fluxes, thereby creating a means for rapid cell fractionation without inflicting cellular damage [Charles et al., Biotechnol. Prog. 23:1463-72 (2007)]. In this study we explored a method to more efficiently select for circulating cancer cells by altering the nanoscale topography of the inner surface of selectin-coated microtubes. This functionalized topography is achieved by attaching naturally occurring halloysite nanotubes to the microtube surface via a monolayer of poly L-lysine, followed by functionalization with recombinant human selectin protein. We have found that halloysite nanotube coatings promote increased capture of circulating cancer cells over a range of selectin surface density while preserving the macroscopic fluid mechanics. Interestingly, at the nano- to micro-scale, the halloysite is found to alter the separation distance between flowing cells and the microtube surface in a predictable way. Ultimately, selectin-functionalized nanotube coatings provide a means for enhanced cancer cell isolation from whole blood.

See more of this Session: Nanomaterials
See more of this Group/Topical: Topical F: Nanotechnology in Medicine

(701c) Functionalizable and Ultra Stable Nanoparticles Coated with Zwitterionic Poly (carboxybetaine) in Undiluted Blood Serum

Wei Yang and Shaoyi Jiang, Chemical Engineering, University of Washington, Seattle, WA

A new surface chemistry presenting an abundance of functional groups for ligand immobilization in an ultra-low fouling background all in one material for nanoparticles was introduced. This surface platform, as demonstrated by zwitterionic poly(carboxybetaine acrylamide) (polyCBAA) and carboxybetaine methacrylate (CBMA) coated nanoparticles, is not only ultra stable in undiluted human blood serum, but also can be conjugated to bio-molecules conveniently and effectively. Thus, this surface chemistry is ideal to create multi-functional nanoparticles for targeted delivery and diagnostics. In addition, this work clearly shows that 10% blood serum commonly used to evaluate the stability of nanoparticles is insufficient and a new evaluation criterion with undiluted blood serum is recommended.

See more of this Session: Nanomaterials
See more of this Group/Topical: Topical F: Nanotechnology in Medicine

(701d) Negatively Charged Gold Nanoparticles Can Inhibit the Formation of Alzheimer's Disease Amyloid-β Protein Aggregates in a Mechanism-Specific Fashion

Deborah Soto-Ortega[1], Stephanie Paolini[1], Alaaldin Alkilany[2], Rahina Mahtab[3], Catherine Murphy[4] and Melissa A. Moss[1], (1)Department of Chemical Engineering, University of South Carolina, Columbia, SC, (2)Department of Chemistry and Biochemistry, University of South Carolina, Columbia, SC, (3)Department of Chemistry, South Carolina State University, Orangeburg, SC, (4)Department of Chemistry, University of Illinois at Urbana-Champaign, Urbana, IL

Alzheimer's disease is a progressive neurodegenerative disorder that has become increasingly prevalent in America's aging population, accounting for 70% of all diagnosed dementia cases. Current therapies for the disease allow AD patients to delay or mitigate the symptoms to improve quality of life but do not stop disease progression. At the pathological level, AD is characterized by the deposition of abnormally aggregated proteins that cause plaque formation. The accumulation of these amyloid plaques within the brain parenchyma is one of the first abnormalities and has been hypothesized to initiate AD pathogenesis; thus, our study addresses amyloid plaque formation.

The primary component of amyloid plaques found in AD is the fibrillar form of the amyloid-β protein (Aβ). During plaque formation, soluble monomeric Aβ, already present in the brain, self-assembles to form soluble aggregates that will further aggregate to create the insoluble fibrils that are deposited. Inhibition of Aβ self-assembly has emerged as a prime therapeutic strategy for this devastating disease. Aβ self-assembly involves multiple reaction intermediates related by nonlinear and interconnected nucleation and growth mechanisms, which provide multiple points for inhibitor intervention.

Nanoparticles have shown promise in biotechnology, including biological imaging and drug delivery. This study evaluates the effect of gold nanoparticles on the overall process of Aβ fibril formation, as well as on the growth of soluble aggregates that occurs later in the self-assembly pathway. Specifically, two growth mechanisms were isolated for study: elongation by monomer deposition and lateral association. The current study examines gold nanospheres ranging in size from 8 to 22 nm that display negatively charged coatings of sodium citrate or positively charged coatings of cetyltrimethylammonium bromide (CTAB) in order to determine the effects of nanoparticle size and charge. Positively charged CTAB coated nanospheres overcoated with negatively charged poly(acrylic acid) (PAA) were also studied to differentiate the effects of nanoparticle coating and charge. All sizes of both sodium citrate coated nanospheres and PAA-overcoated nanospheres inhibited Aβ fibril formation; however, CTAB-coated nanospheres exhibited little effect. These results demonstrate that while nanoparticle size and overcoating have little effect, nanoparticle charge is very important for preventing Aβ fibril formation. Furthermore, these negatively charged nanoparticles selectively reduced the rate of soluble aggregate growth by elongation, indicating that the inhibition of fibril formation occurs in a mechanism-specific fashion.

See more of this Session: Nanomaterials
See more of this Group/Topical: Topical F: Nanotechnology in Medicine

(701e) Production of Robust Virus-Like Particles Via Disulfide-Bond Cross-Linking

Bradley C. Bundy, Chemical Engineering, Brigham Young University, Provo, UT and James R. Swartz, Department of Chemical Engineering, Stanford University, Stanford, CA

Virus-like particles (VLPs) are protein macro-molecular complexes on the order of 10 to 100 nm in diameter and are typically icosahedral in shape and symmetry. VLPs have recently received considerable attention for vaccination, drug delivery, gene therapy and material science applications. VLP-based vaccines have repeatedly demonstrated high responses from low doses due to the strong antigenic nature of their repeated ordered surface. Examples of VLP-based vaccines include the Human Papillomavirus vaccine, Gardasil, approved by the FDA in June 2006 and the Hepatitis B vaccine. By genetically and/or chemically modifying VLPs, VLPs have also been used as efficient cell-specific drug/gene delivery devices and self-assembling nanostructure templates. For the above applications a robust capsid is desired which maintains its structural integrity during storage and while performing its function. One technique used to strengthen VLPs is cross-linking using innate or engineered disulfide bonds. We have assessed the ability to cross-link VLPs with disulfide bonds by either 1) modifying the redox potential during translation and VLP assembly or 2) during a post-assembly incubation for a number of VLPs. Hydrogen Peroxide has been shown to be an effective oxidizer for the formation of disulfide bonds and we also observed significant disulfide bond formation in VLPs by incubation with this oxidizer. However, due to hydrogen peroxide's high standard redox potential of 1.8 volts, non-specific reactions with protein residues which could compromise the integrity of the VLP are a concern. We report the ability to form disulfide-bond cross-linked VLPs by 1) changing the redox potential during protein translation and VLP assembly as well as 2) incubating the assembled VLP with more biologically compatible oxidizers. By optimizing the disulfide-bond cross-linking of VLPs we seek to demonstrate increased VLP stability which will in turn help increase the effectiveness of VLPs as vaccines, drug-delivery agents, and nanostructure templates.

See more of this Session: Nanomaterials
See more of this Group/Topical: Topical F: Nanotechnology in Medicine

Novel Polyurethane/Carbon Nanofiber Composites for Bladder Cancer Applications

Melissa Tsang, Young Wook Chun, and Thomas J. Webster
School of Engineering, Brown University, Providence RI USA

Introduction

Over 38,000 men and 15,000 women are diagnosed with bladder cancer each year in the United States. Treatment often involves cystectomy, or removal of the entire bladder, followed by bladder transplantation. However, organ donations are limited and, as such, synthetic bladder prostheses are appealing alternatives. Polyurethane (PU) is a versatile elastomer that is commonly used in artificial bladders. In turn, materials derived from nanotechnology (that is, those materials possessing one dimension less than 100nm), specifically carbon nanofibers (CNF), have received increasing attention for their potential use in biomedical applications. Recent studies have shown that the dispersion of carbon nanofibers (CNF) in polyurethane (PU) significantly enhanced composite nanoscale surface roughness, tensile properties and thermal stability. Although there have been studies concerning healthy (non-cancerous) cell functions on such nanocomposites, there have been few studies exploring cancer cell responses. Since many patients who receive artificial bladder implants have suffered from bladder cancer and its recurrence poses an imminent risk (especially in the elderly) [1], this study investigated the proliferation and function of both cancerous (specifically, transitional urothelial carcinoma and squamous urothelial carcinoma) and healthy urothelial cells on PU-CNF nanocomposites. As a result, this study created a novel blend of PU-CNF that dually promoted healthy urothelial cell functions and inhibited urothelial carcinoma functions.

Methods

Materials

Polyurethane (PU) and carbon nanofibers (CNF) were dissolved in tetrahydrofuran (THF) and 1,2-dichloroethane, respectively. PU-CNF composites were synthesized by mixing PU and CNF solutions in volumetric ratios of pure PU, 4:1 (PU:CNF), 2:1, 1:1, 1:2, 1:4 and pure CNF by means of ultra-sonication and were vacuum dried on glass discs. Composites were characterized for mechanical properties, wettability, surface roughness and chemical composition by AFM, SEM, XPS, FTIR and goniometry.

Cells

UMUC (ATCC No: CRL-1749) and SCaBER (ATCC No: HTB-3) cells were seeded at 100,000 cells per cm^2 and 82,000 cells per cm^2, respectively, onto all composite types and were allowed to proliferate for 12 hours, 3 days and 7 days, whereas HUC (ScienCell, Cat. No: 4320) were incubated on nanocomposites for 4 hours, 3 days and 5 days. UMUC and SCaBER cells were cultured in minimum essential media supplemented with 10% fetal bovine serum and 1% penicillin/ streptomycin solution. HUC were cultured in urothelial cell medium fortified with 1% urothelial cell growth supplement and a 1% penicillin/ streptomycin solution. Cellular adhesion and proliferation were assessed by MTS assays. Cellular responses were further quantified by measuring the amounts of nuclear mitotic protein 22 (NMP-22), vascular endothelial growth factor (VEGF) and tumor necrosis factor alpha (TNF-alpha). Since bladder cancer cells exhibit elevated levels of NMP-22, measuring cellular release of NMP-22 is a clinical method of diagnosing bladder carcinoma. Similarly, various cancer types, including UMUC and SCaBER cancer cells, demonstrate heightened expression of VEGF, which reflects enhanced cellular growth. In turn, TNF-α is a cytokine involved in inflammation. Dysregulation of TNF-α may be indicative of various forms of cancer.

All tests were conducted in triplicate and numerical data were statistically analyzed using a students' t-test.

Results and Discussion

Addition of CNF Reduced Absorbency and Surface Energy of Nanocomposites

Water contact angle analysis verified that the addition of hydrophobic CNF enhanced the hydrophobicity of the PU-CNF composites. Samples containing equal or more parts of PU than CNF (i.e. pure PU, 4:1 (PU:CNF), 2:1 and 1:1) were more absorbent. Specifically, the absorption of water on these composites was linear with respect to time. **Figure 1** illustrated that, in samples where CNF effects dominated, increasing the amount and, thereby the density, of CNF served to reduce the energy on the composite surface. The trend observed in these results were concurrent with those obtained from water contact angle analysis, since more hydrophilic materials (e.g. samples with more PU than CNF) generally corresponded to higher surface energies.

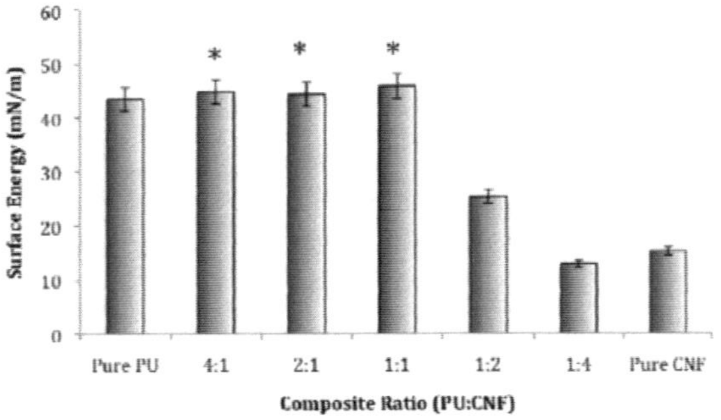

Figure 1. Surface energy of PU-CNF nanocomposites; Data are mean$\pm$SEM; N = 3; * $p<0.05$ as compared to surface energy of nanocomposites with the next greater weight ratio of PU than CNF

Surface Roughness and Surface Stiffness Enhanced with Increasing CNF Density

Figure 2A illustrates the relationship between CNF density and surface roughness of the composite. Increasing CNF density appeared to non-linearly enhance roughness of composite surfaces. The addition of CNF also accentuated the stiffness of PU-CNF composite surfaces, as illustrated in **Figure 2B**.

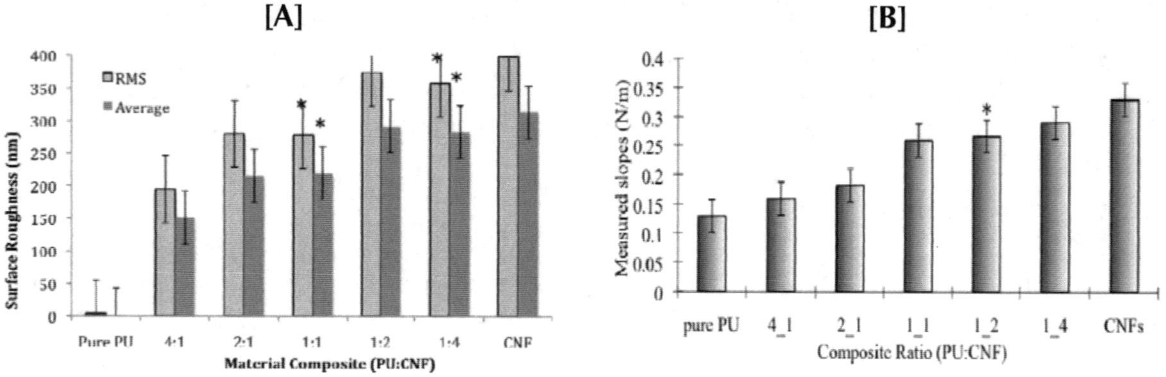

Figure 2. [A] Graph illustrating surface roughness and surface stiffness with respect to the ratio of PU and CNF present in sample, where material composites 1, 2, 3, 4, 5, 6 and 7 correspond to pure PU, 4:1 (PU:CNF), 2:1, 1:1, 1:2, 1:4 and pure CNF, respectively. **[B]** Measurements of nanocomposite surface stiffness; Data are mean±SEM; N = 3; * $p<0.05$ as compared to surface stiffness of nanocomposites with the next greater weight ratio of PU than CNF

CNFs were Evenly Dispersed and Coated with PU

SEM images elucidated the topography of the samples. As shown in **Figure 3**, composites exhibited an even dispersion of CNFs within PU and, in samples with more PU, CNFs appeared to be coated with and deposited in the PU. SEM further verified that increasing the proportion of CNF in composites also increased its density at the surface.

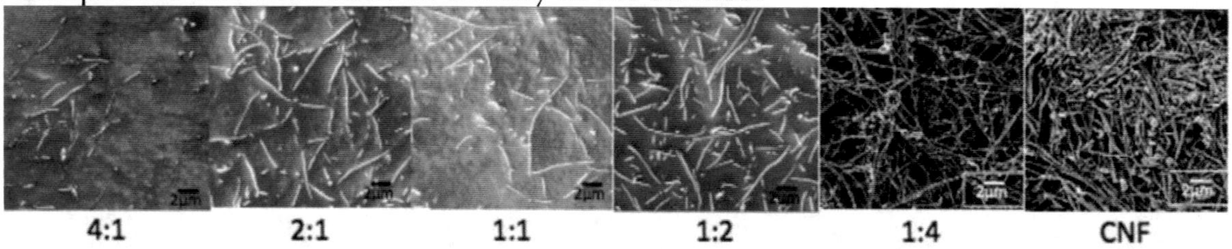

Figure 3. SEM images of 4:1 (PU:CNF), 2:1, 1:1, 1:2, 1:4 and pure CNF composites at 5,000x magnification and 0.50kV

Nanocomposites Exhibited Uniform Chemical Composition

XPS revealed that the chemical composition of all samples (i.e. pure PU, 4:1 (PU:CNF), 2:1, 1:1, 1:2, 1:4 and pure CNF) remained consistent even though they each contained different amounts of PU and CNF. Hence, this study successfully created nanocomposites of varying mechanical properties (e.g. wettability and surface energy) without altering the chemical make-up of its constituents (i.e. PU and CNF). As such, any trends observed in the cellular behavior and activities on these surfaces can be attributed to their mechanical characteristics.

High-Density-CNF Composites Hindered Bladder Cancer Cell Proliferation

Bladder cancer cells responded differently to nanocomposites of varying PU:CNF ratios. Specifically, both UMUC and SCaBER cells demonstrated greater adhesion on composites containing more CNF than PU (i.e. 1:2, 1:4 after 12 hours) (**Figure 4A** and **4B**). However, proliferation rates decreased over time on these same composite samples, thus, providing evidence of a decreased preference of bladder cancer cells to substrates with increased CNFs in PU. This inhibitory effect grew steeper over time. For example, UMUC cell viability dropped

16.18% after 3 days and 31.11% after 7 days from 1:2 (PU:CNF) to 1:4 nanocomposites. In addition, UMUC cell viability significantly diminished from 3 days to 7 days on the same composite type for samples containing more CNF than PU. For example, UMUC cell viability on 1:2 (PU:CNF), 1:4 and pure CNF composites fell by 31.50%, 43.71% and 19.18%, respectively. Similarly, SCaBER cell viability dropped 76.31%, 73.98% and 78.56% on 1:2 (PU:CNF), 1:4 and pure CNF nanocomposites, respectively. Altogether, this demonstrated that composites containing more CNF than PU not only inhibited cancer cell viability, but also hindered the proliferation rate of bladder cancer cells at the material surface.

UMUC exhibited the lowest cell viability on composites with a greater ratio of CNF than PU (i.e. 1:2 (PU:CNF) and 1:4) after 3 days and 7 days of culturing (**Figure 4A**). The same trend was observed in SCaBER cancer cells after 7 days (**Figure 4B**). Cell viability on 1:2 (PU:CNF) composites, for example, was 72.78% lower than on 1:1 composites. Differences in SCaBER cell viability after 3 days were either insignificant or small amongst the various PU:CNF composites. In addition, both UMUC and SCaBER cancer cells demonstrated the lowest proliferation rate on composites with higher CNF density. Comparing cell viability after 3 days and 7 days, UMUC viability fell by 31.50% and 43.71% and SCaBER fell by 76.31% and 74.98% on 1:2 (PU:CNF) and 1:4 composites, respectively. These results implicated that certain PU-CNF composites (i.e. 1:2 and 1:4) dually inhibited the viability as well as the proliferation rate of both UMUC and SCaBER bladder cancer cells.

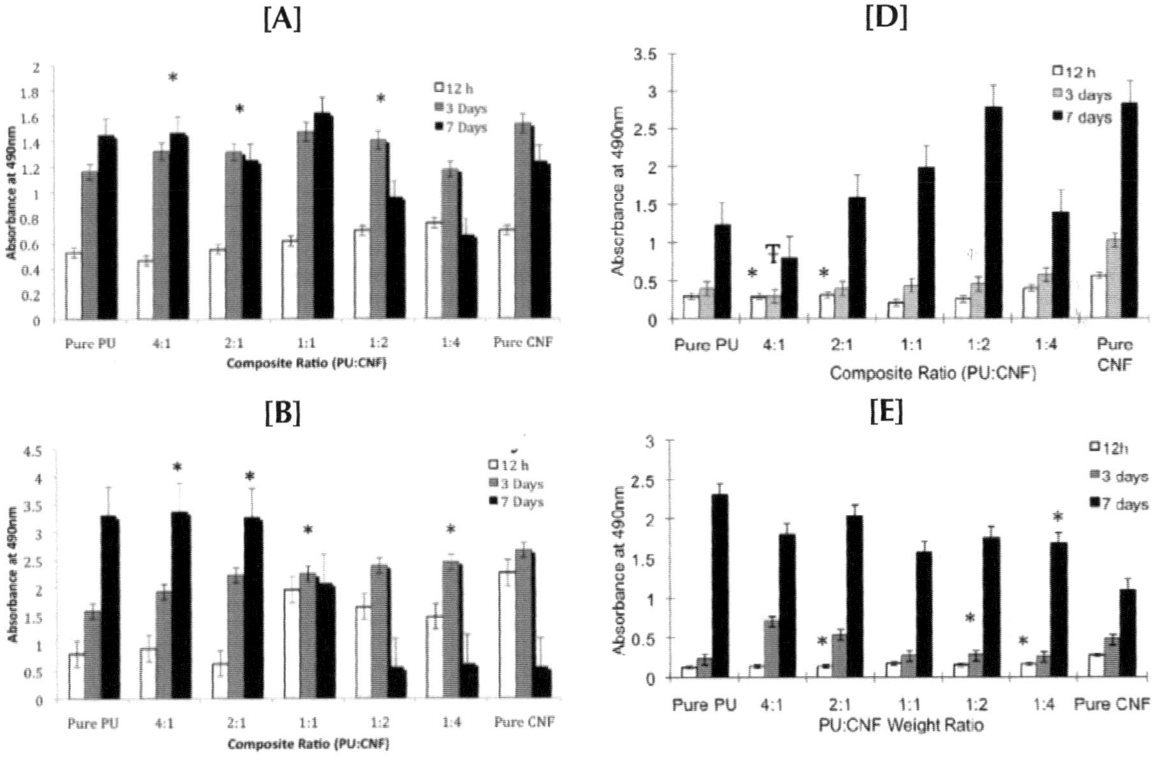

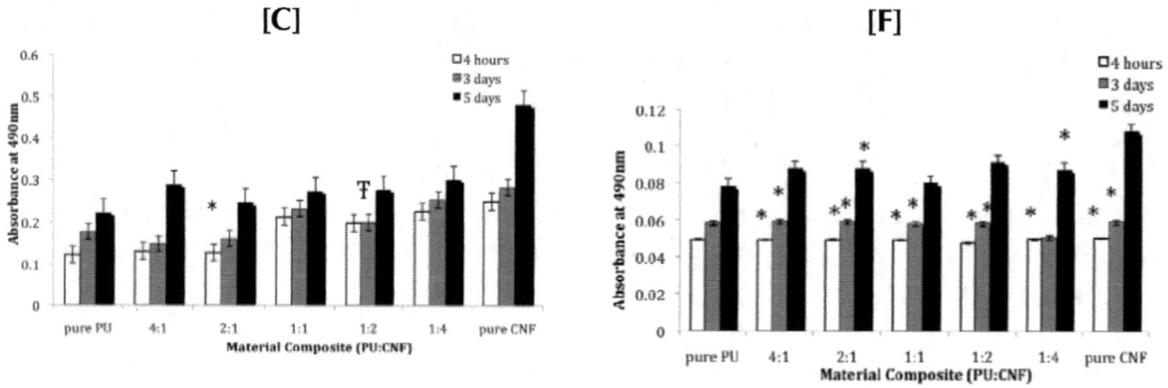

Figure 4. Cell viability of **[A]** UMUC and **[B]** SCaBER cancer cells after 12 hours, 3 days and 7 days on different PU:CNF composites. **[C]** HUC viability after 4 hours, 3 days and 5 days of proliferation on different PU:CNT composites. Results were obtained using the CellTiter 96® AQ$_{ueous}$ One Solution Cell Proliferation Assay (Promega, Part No: TB245) and absorbance was measured using a spectrophotometer at 490nm; Expression of nuclear mitotic protein 22 (NMP-22), a nuclear mitotic apparatus (NuMA) specific to human urothelial cells, in **[D]** UMUC and **[E]** SCaBER bladder cancer cells after 12 hours, 3 days and 7 days on different PU:CNF composites. **[F]** Expression of NMP-22 in HUC healthy bladder cells after 4 hours, 3 days and 5 days. Results were obtained using the Matritech human NMP-22 enzyme immunoassay. Data are mean±SEM; N = 3; * $p<0.05$ as compared to cell viability on nanocomposites with the next greater weight ratio of PU than CNF at the same time point; Ŧ $p<0.05$ as compared to previous time point.

High-Density-CNF Composites Further Reduced Bladder Cancer Cell Expression of NMP-22, TNF-α and VEGF

PU-CNF composites did not induce significant increases in NMP-22 expression in UMUC and SCaBER cells after 12 hours or 3 days (**Figure 4D** and **4E**). In fact, NMP-22 levels remained either less than or comparable to that of HUC until 7 days, suggesting that PU-CNF composites may play a role in delaying the growth and activity of cancerous bladder cells. After 7 days, however, UMUC expression of NMP-22 rose on 1:2 (PU:CNF) composites and, aside from pure CNF samples, elicited the greatest release of NMP-22. In contrast, SCaBER expression of NMP-22 after 7 days decreased as CNF density increased, such that NMP-22 levels were lower on 1:2 (PU:CNF) and 1:4 nanocomposites than on those with greater amounts of PU. It should also be noted that UMUC levels of NMP-22 similarly declined on 1:4 (PU:CNF) composites after 7 days. NMP-22 is a critical protein that facilitates DNA replication, RNA synthesis and hormone binding in bladder cells. Furthermore, levels of NMP-22 in urine are used as a clinical test to diagnose bladder carcinoma. Hence, these trends in UMUC and SCaBER cell expression of NMP-22 suggest that composites with a higher CNF density hinder the mitotic expression of bladder cancer cells.

High-density-CNF composites significantly inhibited the release of TNF-α from both UMUC and SCaBER cancer cells (**Figure 5A** and **5B**). Specifically, TNF-α levels in UMUC after 3 days were 75.22% and 76.08% lower in 1:2 (PU:CNF) and 1:4 composites than on 4:1 samples. For the same time point, SCaBER cell levels of TNF-α in 1:4 (PU:CNF) composites declined by 57.42% as compared to that of pure PU. Overall, the increasing proportion of CNF to PU, or the increasing density of CNF, appeared to deter bladder cancer cell release of TNF-α. TNF-α serves as an immunostimulant and mediates host inflammatory response. TNF-α stimulates production of matrix metalloproteinases (e.g. MMP-2 and MMP-9) in bladder cancer cells [2, 3]. Its presence reflects cancer cell activity. The diminished TNF-α levels observed in UMUC and SCaBER cancer cells on high-density-CNF composites provide strong evidence that these materials exhibit anti-cancerous properties by down regulating inflammatory activity.

VEGF levels in UMUC cancer cells were, on the whole, higher in composites with greater CNF than PU (i.e. 1:1 (PU:CNF), 1:2 and 1:4) after 3 days (**Figure 5C**). Amongst these composites, however, VEGF levels declined as CNF density increased. Thus, UMUC expression of VEGF on 1:2 (PU:CNF) and 1:4 composites were lower than on 1:1, but remained higher than on 4:1 and 2:1 composites. Along a similar vein, VEGF levels in SCaBER cells showed an initial increase from pure PU to 4:1 (PU:CNF) after 3 days, but subsided as CNF density in nanocomposites increased, as seen with 1:4 (PU:CNF) composites (**Figure 5D**). VEGF is a chemical signal released by cells to stimulate the growth of new blood vessels. Since blood vessels supply cells with the necessary nutrients and oxygen to grow, upregulation of VEGF in cancer cells is indicative of their ability to grow and metastasize [4, 5]. Thus, it can be inferred that PU-CNF composites with a high CNF density (i.e. 1:2 (PU:CNF) and 1:4) may be able to decelerate the malignant growth of SCaBER cancer cells. Although these composites were unable to hasten VEGF expression in UMUC cells to the same extent, increasing CNF density did not further elevate VEGF expression. Taken together, high density CNF composites had a positive effect in retard bladder cancer cell growth, with a more pronounced effect on SCaBER cells.

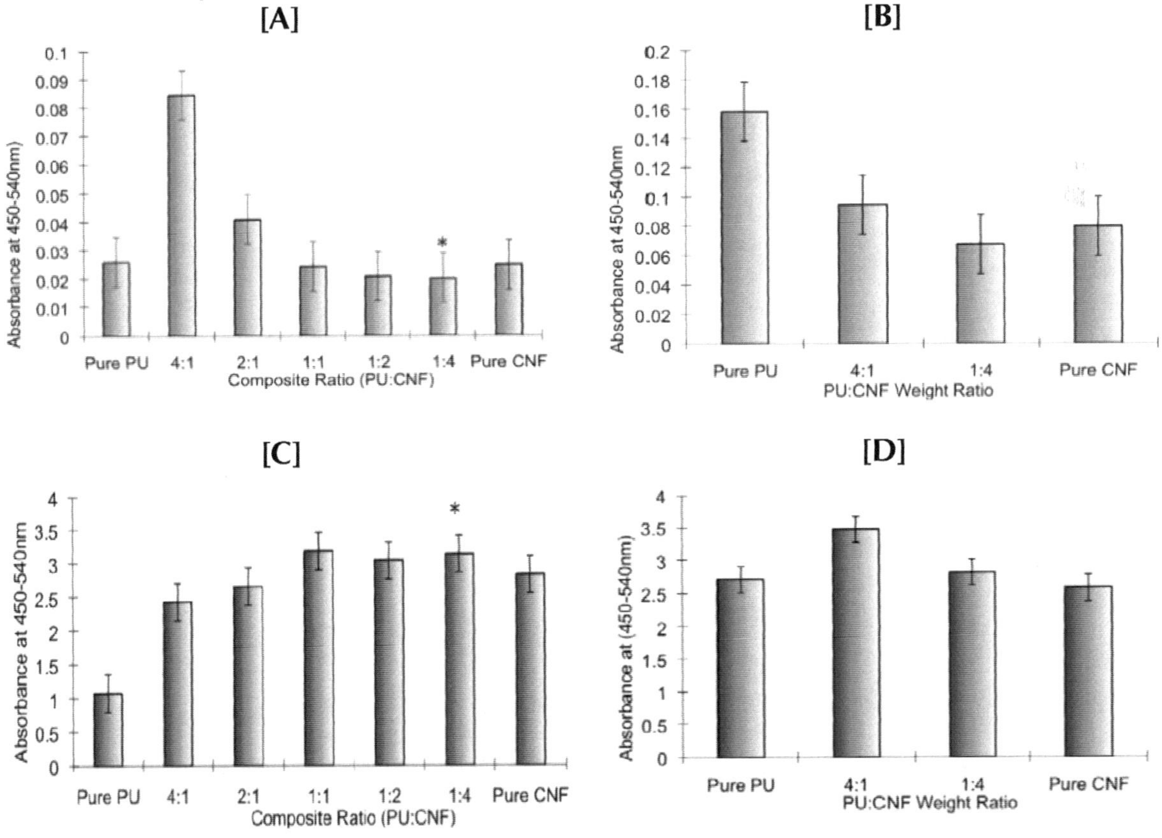

Figure 5. Quantification of the release of TNF-α from **[A]** UMUC and **[B]** SCaBER

bladder cancer cells and VEGF from **[C]** UMUC and **[D]** SCaBER bladder cancer cells after 3 days on various PU:CNF composites. Results were obtained using the R&D Systems Quantikine ELISA kit for mouse TNF-α and human VEGF and measured using a spectrophotometer on a differential setting (450-540nm). Data are mean±SEM; N = 3; * $p<0.05$ as compared to TNF-α expression on nanocomposites with the next greater weight ratio of PU than CNF at the same time point.

Novel PU-CNF Composite for Bladder Prostheses

In addition to monitoring the biological activity of cancerous bladder cells, this study also explored healthy bladder cell function on PU-CNF composites of varying weight ratios. HUC cell viability was linearly proportional to the CNF density in the composite for all time points (**Figure 4C**) and, thereby, established that nanocomposites with a high density of CNF promoted both the adhesion and proliferation of healthy bladder cells. HUC expression of NMP-22 appeared to act independent of PU:CNF ratios (**Figure 4F**). NMP-22 levels were similar amongst all materials after 4 hours and 3 days, but exhibited a slight increase by day 5. Even so, NMP-22 levels remained over 10 times less than that of UMUC and SCaBER cancer cells. These results corroborated cell viability analyses, confirming that PU-CNF composites display no noticeable detrimental effects on healthy bladder cell function; HUC were able to grow and proliferated (at a healthy rate) on these composites.

Whereas PU-CNF composites bore no harmful effects on healthy bladder cells, results established that they posed antagonistic effects on cancerous bladder cells. Specifically, those composites with a greater proportion, or density, of CNF selectively reduced cell viability, diminished proliferation rate, significantly down-regulated levels of TNF-α and, overall, deterred cell growth and activity of UMUC and SCaBER bladder cancer cells. 90% of all bladder cancer originates in the transitional urothelial layer (UMUC) and the remainder mainly lies in the squamous layer (SCaBER). Hence, this study successfully designed a novel PU-CNF composite for bladder prosthetic applications that can dually minimize the chance of cancer recurrence (for both UMUC and SCaBER cases) without hampering healthy bladder cell growth. Moreover, these novel composites achieved such goals by modifying surface properties of the implant alone, without the use of pharmaceutical agents. This study demonstrated that bladder prostheses are a viable and favorable means of bladder replacement. Whereas organ donations are limited, artificial bladders can be easily manufactured. Furthermore, our novel composite offers prophylactic properties to minimize the recurrence and progression of cancer in the bladder.

The basic function of the bladder is to store urine and in normal bladders, this requires tissue elasticity to range from 0.06 N/mm to 0.25 N/mm [6]. Our composites, however, were unable to satisfy this criterion. The authors of this study endeavor to, in future studies, design a bladder prosthetic material that features mechanical properties (i.e. wettability, surface energy and surface roughness) similar to that of 1:2 (PU:CNF) or 1:4 nanocomposites and also provides a practical range of elasticity.

Conclusions

PU is a common biomaterial and studies have shown that PU fortified with CNF boasts superior mechanical and thermal properties. This study determined that healthy and cancerous bladder cells responded selectively to PU-CNF composites of varying weight ratios. Since a majority of patients that require bladder replacements have been victim to bladder cancer, it seemed advantageous to design a prosthetic that will facilitate healthy bladder cell regeneration while minimizing the risk of cancer recurrence. 1:2 (PU:CNF) and 1:4 composites, which featured lower surface energy, enhanced nano surface roughness and uniform dispersion of CNF were able to achieve this goal. Of particular importance is that this novel composite can be easily produced and does not involve any pharmaceutical agents, which may have systemic effects on the body. Although these novel composites elicited optimal biological responses, its success as a bladder prosthetic is contingent on its ability to restore patients with basic bladder function, urine storage. This requires a level of elasticity, however, that the current composite was unable to emulate. As such, future studies should aim to design an improved bladder prosthetic material that features the appropriate biological properties while maintaining the elasticity of native bladder tissue.

Acknowledgements

The authors of this study sincerely thank Dr. Thomas Webster and the Nanomedicine Laboratory at the Brown University School of Engineering for their continued guidance and support of this project. In addition, the authors would like to Dr. Dongwoo Khang for his assistance in performing AFM analysis of the materials.

References

1. Nielsen ME, Shariat S.F., Karakiewicz P.I., Lotan Y., Rogers C.G., Amiel G.E., Bastian P.J., (...), Palapattu G.S. Advanced Age Is Associated with Poorer Bladder Cancer-Specific Survival in Patients Treated with Radical Cystectomy. European Urology 2007;51(3):699-708.
2. Shin KY, Moon HS, Park HY, Lee TY, Woo YN, Kim HJ, Lee SJ, Kong G. Effects of tumor necrosis factor-α and interferon-γ on expressions of matrix metalloproteinase-2 and -9 in human bladder cancer cells. Cancer Letters 2000;159(2)127-134.
3. Lee SJ, Park SS, Lee US, Kim WJ, Moon SK. Signaling pathway for TNF-α-induced MMP-9 expression: Mediation through p38 MAP kinase, and inhibition by anti-cancer molecule magnolol in human urinary bladder cancer 5637 cells. International Immunopharmacology 2008;8(13-14):1821-1826.
4. Förster Y, Meye A, Krause S, Schwenzer B. Antisense-mediated VEGF suppression in bladder and breast cancer cells. Cancer Letters 2004;212(1):95-103.
5. Yang CC, Chu KC, Yeh WM. The expression of vascular endothelial growth factor in transitional cell carcinoma of urinary bladder is correlated with cancer progression. Urologic Oncology: Seminars and Original Investigations 2004;22(1):1-6.
6. Farhat WA, Chen J, Hang J, Antoon R, Litman J, Sherman C, et al. Porcine bladder Acellular matrix (ACM); protein expression, mechanical properties. Biomedical Materials. 2008:3.

(701g) Guanidylamided PEI as Ovarian Cancer Gene Therapy Vectors

Bo Zhang[1], Xinpeng Ma[2], Jianbin Tang[3], Maohong Fan[2], Edward Van Kirk[4], William Murdoch[4] and Youqing Shen[5], (1)University of Wyoming, Laramie, WY, (2)Department of Chemical and Petroleum Engineering, University of Wyoming, Laramie, WY, (3)Department of Chemical and Biochemical Engineering, Zhejiang University, Hangzhou, China, (4)Department of Animal Science, University of Wyoming, Laramie, WY, (5)Department of Chemical and Biochemical Engineering, Zhejiang University, HangZhou, China

A series of guanidylamided linear polyethyleneimine (GPEI) with pendant guanidine groups were synthesized to investigate their gene deliver capability. GPEI showed stronger binding ability with plasmid DNA (pDNA) than linear PEI (lPEI) 25k by heparin competitive assay and EB exclusion assay, forming small size nanoparticles (100 - 150 nm) with zeta potential ranging from 15 to 30 mV. The transfection efficiency in two ovarian cancer cell lines was determined. GPEI had improved transfection efficiency than its corresponding linear PEI. Meanwhile, the GPEI was much less toxic than the original PEI 10k (lPEI230) in both cell lines tested. GPEI penetrated cellular membrane more efficiently than LPEI 25k but in cytosol its disassociation with pDNA was much slower.

See more of this Session: Nanomaterials
See more of this Group/Topical: Topical F: Nanotechnology in Medicine

(725a) Fabrication of Nanobiosensors

Ajit Sadana, Chemical Engineering Department, University of Mississippi, University, MS

Different nanobiosensor fabrication methods are presented and reviewed. The examples of nanobiosensor fabrication presented are selected at random from the literature. These examples have appeared recently in the open literature. Though glucose nanobiosensor fabrication methods presented appear frequently in the literature, other examples of nanobiosensor fabrication are also presented. The nanobiosensor fabrication methods for glucose sensors include: biosensor using a nanocomposite electrode, Safavi et al., 2009; screen-printed water-based carbon-ink microband biosensor; Pemberton et al., 2009; zinc-oxide nanoparticle/glucose oxidase biosensor, Ren et al., 2009; and platinum nanowire array biosensor, Yang et al., 2006). Other fabrication techniques presented include: biosensor fabrication procedure based on processable polyanilione nanoparticles, Morrin et al., 2005; biosensor fabrication based on nanowire microelectrode array, Yang et al., 2006; and recrystallization technologies to fabricate a Si nanowire biosensor, Ashburn et al., 2009). The presentaton of different nanobiosensor fabrication techniques together quickly provides one with an overall perspective of the different fabrication techniques that have appeared in the recent literature, and what techniques exhibit potential, and which techniques are being explored further research wise. The nanofabrication techniques are application dependent, such as the determination of glucose, but as one might reasonably expect, biosensors for the detecton of other analyztes is slowly gaining importance, and thus nanofabrication techniques other than for glucose detection need to be explored further. In a sense, the nanobiosensor fabrication btechniques briefly mentioned here may be considerd loosely as a starting point in an iterative solution, which continually improves with each iteration.

See more of this Session: Diagnostics and Nanoimaging
See more of this Group/Topical: Topical F: Nanotechnology in Medicine

(725b) Size Control for Fluorescent Polymeric Nanosensors

Kevin J. Cash, J. Matthew Dubach, Mary K. Balaconis and Heather A. Clark, Biomedical Engineering Group, The Charles Stark Draper Laboratory, Cambridge, MA

Recent years have shown the incredible versatility and utility of polymeric nanosensors. Sensors have been developed for a wide variety of ionic analytes (e.g. sodium, potassium, pH) as well as nonionic analytics (e.g. glucose) and have been applied for challenging research fields such as detection of sodium sparks in live cardiomyocytes. This field of research has yielded new knowledge on cellular metabolism and function. Additionally, these sensors can be applied in vivo, through implantation in or under the skin in animals; showing potential as a research tool for continuous monitoring of analyte concentrations without the need for direct samples to be taken from the animal.

The size control of these nanosensors is an important parameter in successful applications. For in vitro research in cell lines, the use of smaller sensors is preferred for improved distribution throughout the cell as well as to minimize clogging of the micropipette tip during injection. However, for in vivo applications, larger sensors are desirable in order to minimize sensor migration away from the injection site. Methods to affect sensor size over both a small and large range will be discussed for both in vitro and in vivo applications.

This work was supported by a grant from the National Institutes of Health National Institute of General Medical Sciences (R01 GM084366).

See more of this Session: Diagnostics and Nanoimaging
See more of this Group/Topical: Topical F: Nanotechnology in Medicine

(725c) Nanobiosensor Applications

Ajit Sadana, Chemical Engineering Department, University of Mississippi, University, MS

Different examples of nanobiosensors that have recently appeared in the open literature are briefly reviewed and analyzed. The examples were selected at random and may be classified into three broad categories depending on the analyte detected. These categories are: (a) medical applications, (b) DNA/RNA applications, and (c) other applications. The examples presented together and briefly analyzed provide one with a perspective of the different nanotechnology techniques and processes that have recently been used to develop effective nanobiosensors. The brief clasification presented above may not be a 'true picture' of the different categories where nanobiosensors are presently being used. Many more examples need to be included. Perhaps, other categories need to be defined where the classification may be expanded to include other examples. Neverthelees, with the classification presented the predominance of medical applications is clearly prevalent. With the current emphasis on the detection of biomarkers for different diseases such as MI (myocardial infarction), and autoimmune diseases such as RA (rheumatoid arthritis), and SLE (systemic lupus erythematosus), medical applications of biosensors will continue to dominate the market, and the application of nanobiosensors to this area is not to be entiely unexpected. Of course, the quantitative detection of glucose in blood to manage DM (diabetes mellitus) will, as expected, continue to dominate the biosensor market thanks to the increasing obesity and minimum, if any, time spent by individuals on exercising activities. Many more biosensor applications selected preferably at random need to be picked up and placed in approporiate categories. Nevertheless, if one were to be cautious, one may state that it seems that applications of biosensors are expanding especially in the non-medical category. This is not entirely unexpected due to the ease of applications of biosensors to different areas and fields. This is reinforced by papers that were presented at two very recent international conferences on biosensor applications.

See more of this Session: Diagnostics and Nanoimaging
See more of this Group/Topical: Topical F: Nanotechnology in Medicine

(725d) Resolving Subdiffraction Limit Distances Using Plasmon Coupling Microscopy

Bjoern Reinhard, Chemistry, Boston University, Boston, MA

Discrete assemblies of noble metal nanoparticles enable new imaging and improved sensing applications. The distance dependent interactions between individual noble metal nanoparticles can be used to engineer discrete plasmonic nanostructures for applications in microscopy, sensing and imaging. For instance, DNA or RNA tethered noble metal nanoparticles (so called plasmon rulers) are active nanostructures which can indicate nanoscale distance changes through shifts in their resonance wavelength. Plasmon coupling measurements are not restricted to monitoring end-to-end distances of biopolymers in plasmon rulers. Instead, plasmon coupling can also resolve subdiffraction limit contacts between individual nanoparticle labeled surface receptors that perform a lateral diffusion on a cell surface. If two nanoparticles approach each other close enough for their plasmons to couple, the resonance wavelength of the dimer red-shifts. This effect is utilized in plasmon coupling microscopy (PCM). Individual, independent particles can be tracked with high spatial and temporal resolution in PCM. If two particles colocalize and can no longer be discerned optically, PCM monitors changes in the plasmon resonance wavelength to detect direct contacts between individual particles. Spectral shifts in the plasmon resonance of diffusing nanoparticles during optical colocalization can be detected in an optical widefield microscope using a ratiometric imaging approach. The sample is illuminated with unpolarized whitelight in conventional darkfield microscopy, and the detected image is split into two monochromatic images. Spectral shifts in the resonance wavelengths of two colocalized particles can then be detected as a change in the ratio of the dimer image in the two channels. The technology enables, for instance, to resolve close contacts between nanoparticle labeled cell surface receptors.

See more of this Session: Diagnostics and Nanoimaging
See more of this Group/Topical: Topical F: Nanotechnology in Medicine

(725e) Optical Sensing of Acetone in Exhaled Breath Utilizing Acid Catalyst Membranes

Adam D. Worrall[1], Jonathan A. Bernstein[2] and **Anastasios Angelopoulos**[1], (1)Chemical and Materials Engineering, University of Cincinnati, Cincinnati, OH, (2)Division of Immunology, University of Cincinnati Medical School, Cincinnati, OH

Measurement of acetone content in human breath has been suggested as a non-invasive method to the diagnosis of diabetes and/or ketoacidosis. Existing methods employ time consuming, ex-situ analyses of breath condensate that limit broad, real-time application. In the present study, the diffusion and immobilization of dilute fluorescent dyes into the heterogeneous nanophases of perfluorosulfonic acid (PSA) polymer membranes is investigated for direct, one-step gas phase acetone analysis utilizing a hand-held device. The membranes act as acid catalysts for the condensation reaction of acetone and the immobilized dyes at temperatures above the acetone boiling point, yielding an optical response in the visible region of the electromagnetic spectrum that may be observed utilizing standard light emitting diodes (LED's). Partitioning of the dye concentration among the lipophilic, hydrophilic, and transitional nanophases within the PSA membrane is shown to control the sensitivity of the membrane to acetone exposure, down to ppb levels. UV/VIS and FTIR are utilized to quantify the diffusion of the dyes and acetone in the membrane. The use of deuterated acetone as a marker molecule allows for the determination of the intermediate and product molecules (flavans) immobilized within the membranes and to confirm the selectivity of the optical response.

See more of this Session: Diagnostics and Nanoimaging
See more of this Group/Topical: Topical F: Nanotechnology in Medicine

(725f) Transverse Relaxivities of Polyether-Magnetite Complexes: The Effect of Polymer Loading and Composition

Richey M. Davis[1], Matthew R.J. Carroll[2], P.P. Huffstetler[3], William C. Miles[4], J.D. Goff[3], J.S. Riffle[3], Robert C. Woodward[2] and Timothy G. St. Pierre[2], (1)Virginia Tech, Blacksburg, VA, (2)Physics, The University of Western Australia, Crawley, Australia, (3)Macromolecular Science and Engineering, Virginia Tech, Blacksburg, VA, (4)Chemical Engineering, Virginia Tech, Blacksburg, VA

Magnetite nanoparticles have many uses, in particular as T2 contrast agents for magnetic resonance imaging (MRI) and for incorporation into magnetic ferrofluids.[1,2] Magnetite particles have relatively low cytotoxicities, and their surfaces are amenable to the adsorption of biocompatible macromolecules.[3] Thus, specific functionalities can be attached to magnetite nanoparticles in order to optimize a specific property for appropriate applications.[4] This study concerns the transverse relaxivities of polyether-magnetite complexes in which both the magnetite and the polyethers were well-defined. The magnetite was synthesized using a method adapted from Pinna et al.[5], which consists of the high temperature decomposition of iron acetylacetonate at high temperature in benzyl alcohol, resulting in well-defined magnetite nanoparticles. Well-defined polyethers with tri-phosphonate anchor groups were synthesized including poly(ethylene oxide) (PEO) homopolymers and amphiphilic poly(propylene oxide)-b-poly(ethylene oxide) (PPO-b-PEO) block copolymers. The triphosphonate end groups provide robust anchoring of the polyether chains on magnetite in phosphate buffered saline.[6] The number average size of dispersed complexes and the half-life for doublet formation of well-defined polyether-magnetite complexes like those in the present study can be predicted with a model based on a star polymer.[4] In the present study, we report transverse relaxivities of the polyether-magnetite complexes that were measured while varying polymer loading, polymer type, and, in the case of the block copolymers, polymer composition.

1. Huffstetler, P. P, et al. Polymer Preprint, 2008, 49(2), 1103-1104. 2. Mefford, et al. Chemistry of Materials, 2008, 20(6), 2184-2191. 3. Harris, L., et al. Chemistry of Materials 2003, 15(6), 1367-1377. 4. Miles, W.C., et al. Langmuir 2009, 25(2), 803-813. 5. Pinna, N., et al. Chemistry of Materials 2005, 17, 3044-3049. 6. Goff, J.D., et al. Chemistry of Materials 2009, 21, 4784–4795.

See more of this Session: Diagnostics and Nanoimaging
See more of this Group/Topical: Topical F: Nanotechnology in Medicine

AUTHOR INDEX

AUTHOR INDEX